燒味

傳承滋味

Authentic Flavours of
Barbecue Meat

Rachel Yau

　　認識陳永瀚師傅，正值教室需要找一位專業的燒臘師傅，為我們教授一般人認為難以在家做得出色的燒味菜式，例如叉燒、乳豬、燒鵝等。那時廖教賢師傅開始在教室教授中菜，課堂的菜式非常受歡迎，廖師傅有一位好拍檔，聽說燒味做得非常出色，於是他介紹了陳師傅來，為我們提供燒味課程。

　　陳師傅非常專業及細心，當初我們商議了幾道菜式在課堂教授，認真的陳師傅重新鑽研，將幾道本來只在酒樓烹調的菜式，透過家裏的烤爐也可做出來！過了不久，陳師傅教授的桂花汁叉燒旋即成為最熱門的烹飪班，一下子開辦了四十多班，慕名而來的學生來自五湖四海，如法國、瑞士、德國、英國、美加、日本及澳洲，吃過後也大讚是最好吃的叉燒！

　　陳師傅是我最喜歡及最尊敬的中菜師傅之一，為人處事認真，每次上課都盡心竭力把每道菜式達至完美效果！我誠意推介陳師傅最新力作《燒味・傳承滋味》，讓你將即將失傳的經典燒味珍饈承傳下去！

廖 教 賢

　　自從與陳永瀚師傅合作編著食譜至今已兩年多,因各自忙於工作而鮮有合作機會。近聞陳師傅將有新作推出,當然感到十分雀躍,希望有機會拜讀陳師傅之作品。

　　陳師傅做事作風一絲不苟,對菜餚製作認真、細緻,具有個人風格,喜將新舊製作方法及食材互相融合,既可傳承燒味菜餚之精髓,又令菜餚有另一番新面貌。

　　我想陳師傅新著作內的菜式及製作方法,將會帶給讀者們耳目一新之感!

許美德

重新認識香港昔日滋味！

香港向來有「亞洲美食之都」的美譽，琳琅滿目的中式美食之中，除了點心及小炒外，燒味是其中一個大範疇，油雞、滷味、金錢雞等，五花八門。平凡至街邊燒臘檔「斬嚫大叉燒」，甚至是酒樓或酒店味部晚飯的「明爐脆皮燒鵝」，或是宴席喜慶的「黃金乳豬全體」等，都是百多年來經典粵菜的佼佼者。

喜見功力十足的酒店名廚陳永瀚師傅又一傑作，功力深厚的他任職酒店味部掌爐，閒時開班授徒，據我所知報讀他的課堂也非易事，名額經常爆滿，還要抽籤才能成為幸運兒。他把上堂的資料及經驗編著成書——《燒味·傳承滋味》，內容資料豐富準確，將傳統懷舊菜、創新小菜及宴席菜式，沿用「傳統不守舊、創新不忘本」的精神，為讀者提供一本教材式的書籍。

我非常喜歡美食，也尊重「正宗」這兩字，但在平衡創新與傳統之間的關係，此書著實令人著迷，我推薦《燒味·傳承滋味》一書，希望各位讀者喜歡及感受陳師傅的誠意、心機及努力。

自 序

　　燒味菜式是香港飲食文化的一部份，無論大型宴客或家常便飯，經常是餐桌上的主角之一，但有些面臨失傳的菜式，或是創新的佳餚，並不能隨意購買或品嘗得到，更別説怎樣瞭解烹製的方法了。有見及此，我今次特意挑選了很多具代表性的菜式編撰食譜內，並趣談當中的由來及典故，整個烹製過程變為簡易的家庭式做法，利用家居的爐具，令一般人在家也能煮出各類經典燒味菜式，只需自己動手，絕對毋須假手於人。

　　本書分為五個章節——特色冷盤、惹味小食、懷舊燒味、滋味海鮮、歡聚宴客，當中更加插了幾道受歡迎的福建閩南小食及川味菜式，讓大家享受多元化的味覺效果。

　　在此，多謝圓方出版社全人、廖教賢老師、Rachel 老師及許美德師傅，在製作時的協助及寶貴意見。

目　錄 contents

燒 味 汁

帶清香味的豉油汁,用於各類滷水或燒烤菜式,作為添加味道之用,也是碟頭飯的最佳配汁,故又稱為「飯汁」。

○ 料頭

油	1 湯匙
葱段	5 克
薑碎	5 克
乾葱碎	5 克
蒜肉碎	5 克
八角	1 粒
花椒粒	1 茶匙
花雕酒	1 茶匙(後下)

○ 調味料

清水	250 克
老抽	15 克
生抽	50 克
蠔油	30 克
雞粉	1 茶匙
冰糖	50 克
甘草	2 片

做法

燒熱油鑊,下料頭爆香,加入調味料煮滾,用慢火煮 5 分鐘,熄火焗 15 分鐘,隔渣備用。

Homemade barbecue sauce

(for enhancing the flavours in marinated dishes or barbecue meats)

Aromatics:
  1 tbsp oil
  5 g spring onion (cut into short lengths)
  5 g ginger (finely chopped)
  5 g shallot (finely chopped)
  5 g grated garlic
  1 whole pod star anise
  1 tsp Sichuan peppercorns
  1 tsp premium Shaoxing wine (added last)

Sauce base:
  250 g water
  15 g dark soy sauce
  50 g light soy sauce
  30 g oyster sauce
  1 tsp chicken bouillon powder
  50 g rock sugar
  2 slices liquorice

Method:
  Heat oil in a wok. Stir-fry aromatics until fragrant. Add sauce base and bring to the boil. Cook over low heat for 5 minutes. Turn off the heat. Cover the lid and leave it for 15 minutes. Strain and ready to use.

回　鍋　醬

由回鍋肉的烹調方法改創而成，以臘肉代替新鮮五花腩，再配合特色調味料，令味道更香濃，可作為蘸汁或調味料使用。

○材料

油	3 湯匙	老干媽油辣椒	1 湯匙
乾葱茸	30 克	花雕酒	3 湯匙
蒜茸	20 克	甜麵醬	5 湯匙
臘肉	150 克	花椒油	1 湯匙
豆瓣醬	半湯匙		

做法

1. 臘肉隔水蒸 40 分鐘，取出，切成米粒狀備用。

2. 燒熱鍋放入油，加入乾葱茸及蒜茸，用慢火炒至金黃色，下臘肉及其餘調味料拌勻，煮滾後熄火即成。

Twice-cooked sauce
(use as a dipping sauce or seasoning)

Ingredients:
3 tbsp oil
30 g finely chopped shallot
20 g grated garlic
150 g preserved pork belly
1/2 tbsp spicy bean sauce
1 tbsp Lao Gan Ma brand chilli oil
3 tbsp premium Shaoxing wine
5 tbsp sweet soybean paste
1 tbsp Sichuan pepper oil

Method:
1. Steam the preserved pork belly for 40 minutes. Let cool and dice finely.
2. Heat oil in a wok. Stir-fry shallot and garlic over low heat ntil golden. Add preserved pork belly from step 1 and remaining ingredients. Stir well. Bring to the boil and turn off the heat. Ready to use.

回味醬

由多種醬料及調味料調配而成，香味濃郁，經常作為醃料或調味料使用。

◯ 材料

油	2 湯匙	花雕酒	12 克
海鮮醬	60 克	玫瑰露酒	12 克
磨豉醬	60 克	南乳	10 克
蠔油	20 克	砂糖	140 克
生抽	60 克	芝麻醬	40 克

做法

1. 將油及芝麻醬分別放入碗內；其他材料混和備用。

2. 用油起鑊，加入已混和的調味料，用慢火煮滾，最後慢慢倒入芝麻醬拌勻，煮至翻滾即成。

Unforgettable sauce

(for marinating or use as a seasoning)

Ingredients:
2 tbsp oil
60 g Hoi Sin sauce
60 g ground bean sauce
20 g oyster sauce
60 g light soy sauce
12 g premium Shaoxing wine
12 g Chinese rose wine
10 g fermented red tarocurd
140 g sugar
40 g sesame paste

Method:
1. Set aside the oil and the sesame paste separately. Mix the rest of the ingredients.
2. Heat a wok and add the oil. Pour in the sauce mix from step 1. Bring to the boil over low heat. Stir in the sesame paste slowly at last. Bring to the boil. Ready to use.

醃味
白滷水汁

由多種香料藥材煲製而成，用以醃浸各類食材調味之用，可生醃或煮熟後醃浸，用途廣泛，如海鮮類、家禽類、家畜類也很合適。

○材料

清水 600 克 / 沙薑片 10 粒 /
丁香 1 茶匙 / 香葉 5 片 / 小茴香 1 湯匙 /
甘草 3 片 / 白豆蔻 10 粒 /
胡椒粉半茶匙 / 薑片 15 克 / 南薑片 15 克

○調味料

鹽 2 湯匙 / 雞粉 1 湯匙

做法

材料煮滾，轉慢火加蓋煲 30 分鐘，加入調味料攪拌至鹽完全溶化，熄火待冷備用。

Spiced white marinade
(for marinating seafood, pourlty or meat)

Ingredients:
  600 g water
  10 sand ginger roots
  1 tsp clove
  5 bay leaves
  1 tbsp cumin
  3 slices liquorice
  10 white cardamom pods
  1/2 tsp ground white pepper
  15 g sliced ginger
  15 g sliced galangal

Seasoning:
  2 tbsp salt
  1 tbsp chicken bouillon powder

Method:
  Boil all ingredients. Turn to low heat and cook for 30 minutes. Add seasoning and stir until salt dissolves. Turn off the heat and let cool.

A 類上皮水

由於糖分含量較少，適合用於偏高溫油炸的菜式。貯存於雪櫃保鮮，可保持最佳效果。

○材料

清水 75 克 / 麥芽糖 20 克 /
白醋 80 克 / 大紅浙醋 5 克 /
雙蒸酒 45 克 / 玫瑰露酒 5 克

做法

清水與麥芽糖混和，座於熱水內至麥芽糖完全溶化，加入白醋及大紅浙醋，待冷卻後，加入雙蒸酒及玫瑰露酒拌勻即成。

Basting sauce A

Ingredients:
  75 g water
  20 g maltose
  80 g white vinegar
  5 g red vinegar
  45 g double-distilled rice wine
  5 g Chinese rose wine

Method:
  Mix water with maltose. Put it over a pot of simmering water and stir until maltose dissolves. Add white vinegar and red vinegar. Wait till it cools. Stir in rice wine and Chinese rose wine.

認 識 燒 味 工 具
Common utensils for barbecue meats

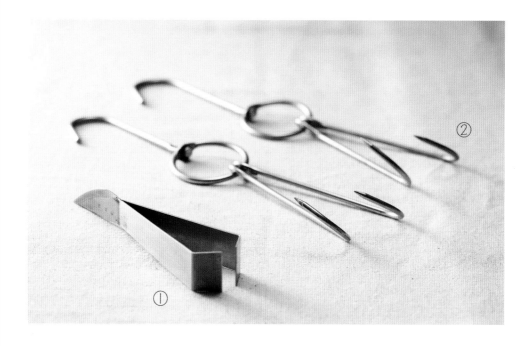

① 鋼 毛 刀 （鴨 毛 鉗）
Feather tweezer

拔除短毛及細毛。

It is for plucking short and fine feathers from birds.

② 鴨 環
Duck hook

雙勾設計，用以勾掛雞鴨家禽，風乾或塗抹時使用。

It is a double-sided hook for hanging poultry like chicken or duck. Use it when air-drying the poultry or brushing on basting sauce.

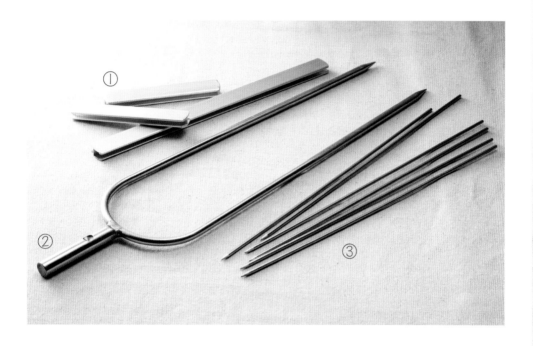

① 鋁合金乳豬柴
Aluminium roast pig planks

烤製乳豬時，放入腹腔撐開，有長短不同尺吋。

This is for spreading and flattening the belly of a suckling pig before roasting. It comes in different sizes.

② 不銹鋼乳豬叉
Stainless steel fork

大型的乳豬叉，串着乳豬固定位置，再進行烤製，有大小之別。

It is a big fork for trussing and securing a suckling pig before roasting. It also comes in different sizes.

③ 不銹鋼叉燒針
Feather tweezer

長形的刺針，串掛叉燒、金錢雞或固定乳豬烤製之用，用途廣泛。

It is a long metal skewer for hanging barbecue pork, chicken liver and pork medallions and fixing the suckling pig while roasting.

用醬油、調味、滷水拌醃的冷盤，
蘊藏的滋味，
令你回味、再回味！

Cold appetizers

Tossed in soy sauce, seasoning and spiced marinade
The depth and complexity in taste
Simply unforgettable

話　梅
花　雕　醉　鴿

Drunken squab in
Shaoxing wine with
dried plums

6人份量

○材料

乳鴿	約 450 克

○話梅酒材料

花雕酒	120 克
甜話梅	2 粒
糖桂花	半湯匙

○滷水料

清水	900 克
八角	2 粒
香葉	4 片
沙薑	3 粒
小茴香	1 湯匙
薑片	15 克
甜話梅	4 粒

○滷水調味料

幼鹽	70 克
雞粉	30 克
花雕酒	80 克

做法

1. 預先將話梅酒材料混合，浸泡兩天，使話梅味滲透花雕酒，備用。

2. 煲滾滷水料，轉慢火煮10分鐘，熄火，待冷後，加入調味料拌勻至幼鹽完全溶化，備用。

3. 乳鴿掏去肥膏及內臟，洗淨。

4. 燒熱水，放入乳鴿至水滾，轉慢火煮5分鐘，熄火，加蓋焗30分鐘至熟透。將乳鴿放入滷水內浸3小時。

5. 享用時，將乳鴿斬件，淋上話梅酒即可。

必 學 重 點

乳鴿放入滷水內浸泡時間足夠，酒味更濃郁。
若時間充裕泡浸一天，風味更佳。

棒 棒 雞

Bang Bang chicken

此菜起源於四川漢陽鎮，採用當地飼養的雞，由於雞隻在花生沙地上放養，故肉質細嫩、肥美。

斬雞時，必須使用木棒協助敲打刀背，才容易斬開雞身，去骨後，再用木棒捶鬆雞肉，切成粗條，故得名為「棒棒雞」。

現今，此菜式多會加入粉皮、萵筍、小黃瓜或沙律菜等拌食，增添風味。

6 人份量

◯ 材料

光雞	半隻（約 700 克）
鮮粉皮	600 克
青瓜	150 克
大葱	30 克

◯ 醬汁

砂糖	2 茶匙	生抽	2 湯匙
幼鹽	半茶匙	芝麻醬	2 湯匙
雞粉	半茶匙	老干媽油辣椒	2 湯匙
花椒油	1 茶匙	蒜茸	2 湯匙
鎮江醋	4 茶匙	芫茜段	6 湯匙
麻油	2 湯匙		

做法

1. 將醬汁攪拌均勻，備用。

2. 光雞去除內臟，洗淨。

3. 燒熱水，放入光雞至水滾，轉慢火煮 5 分鐘，熄火，
   加蓋焗 30 分鐘至熟透，取出，放於冷開水待涼。

4. 粉皮切成約半吋寬條狀，放入滾水燙至透明，盛
   起，泡冷開水至冷卻，瀝乾水分備用。

5. 雞去骨，將雞肉、青瓜及大葱切成幼條，放入碟內，
   拌入粉皮及醬汁即可。

必 學 重 點

燙粉皮要掌握時間，當見呈現透明狀即撈起，
以免過熟收縮及太軟身，影響口感。

○ 材料

溫室小青瓜　　　900 克

○ 醃汁

山西老陳醋	70 克	炸蒜茸	2 湯匙
生抽	10 克	砂糖	50 克
麻油	30 克	幼鹽	10 克
豆瓣醬	30 克	雞粉	10 克
蒜肉	50 克（拍碎）	花椒辣油	1 茶匙
芫茜	10 克（切段）		

做法

1. 小青瓜洗淨，瀝乾水分，切去頭尾兩端，用刀拍扁
   至裂開，去籽，再切成約 2 吋半長瓜條，備用。
2. 醃汁混合後，加入青瓜條拌勻醃 1 小時即成。

必 學 重 點

- 醃汁建議預早一小時混合，使蒜肉充分地
  滲透香味。
- 青瓜條放入醃汁後，宜每隔十五分鐘攪拌
  一次，使青瓜條更均勻入味。

老醋
蒜香瓜

Pickled cucumber in
aged vinegar and
garlic dressing

爽脆
酸子薑

Pickled young ginger

6 人份量

◯材料　　　　　　　　◯醃汁

| 子薑 | 1.2 千克 |
| 幼鹽 | 80 克 |

清水	900 克
冰糖	900 克
幼鹽	30 克
潮汕米醋	600 克
檸檬	3 片

必 學 重 點

■ 若以白醋代替米醋，必須留意醃汁混合後
的酸度，若味道太酸會令薑肉本身的水分
流失，令子薑收縮及皮韌。 若真的太酸，
只需適當地增加清水及冰糖的份量，中和
即可。

■ 應放於雪櫃貯存，保鮮之餘，口感也佳。

做法

1. 煲滾清水 900 克，轉慢火，加入冰糖及幼鹽攪拌至完全溶化，熄火，待冷，加入潮汕米醋及檸檬拌勻，備用。

2. 子薑洗淨，去皮，切成筷子般厚度的片狀，灑入幼鹽 80 克拌勻醃 45 分鐘，用清水沖洗約 90 分鐘至鹹味完全消失。

3. 用毛巾抹乾子薑表面水分，放入醃汁內，用重物輕壓浮面的薑片，確保薑片完全浸泡於醃汁，需浸泡兩天或以上。

冰 鎮
鮮 鮑 魚

Chilled marinated abalones

○ 材料

鮮鮑魚仔	6 隻
薑片	20 克
葱	20 克

○ 滷水汁

清水	300 毫升	沙薑	2 片
南薑	10 克	甘草	1 片
香茅	1 支	陳皮	少許
薑	10 克	紅尖椒	半隻
香葉	2 片		

○ 滷水調味料

冰糖	18 克
幼鹽	9 克
雞粉	2 克

做法

1. 將食鹽 1 茶匙加入 50℃溫水（約 300 毫升）拌勻，放入鮮鮑魚浸泡約 10 分鐘，取出洗淨，用清水沖洗數分鐘。

2. 滷水汁煮滾，加蓋，用慢火煮約 8 分鐘，加入調味料攪拌至溶化，熄火後待冷。

3. 燒熱水至 50℃，加入薑片、葱及鮮鮑魚，慢火煮滾後，再煮約 2 分鐘，熟透後取出鮑魚肉，去除內臟，洗淨。

4. 鮑魚肉抹乾水分，放入滷水汁浸約 5 小時，冷吃品嘗。

必學重點

- 先將鮑魚浸於有溫度的鹽水內，污垢會慢慢浮起，並使鮑魚肉脹身，易於洗擦。
- 開始煮鮑魚時，水溫不宜太高，以免鮑魚肉收縮太快，影響外觀及口感。

山葵醬
手撕雞

Hand-shredded chicken dressed in
wasabi sauce

6 人份量

○材料		○調味料	
雞胸肉	1 個	蠔油	4 茶匙
	（約 200 克）	砂糖	4.5 茶匙
西芹	40 克	麻油	1 茶匙
甘筍	40 克	雞粉	1/4 茶匙
甜紅椒	30 克	老抽	半茶匙
甜黃椒	30 克	山葵醬（青芥辣）	1 茶匙
京葱	20 克		
酸子薑	20 克		
炒熟芝麻	1 茶匙		
飛魚子	1 湯匙		

做法

1. 燒熱水，放入雞胸肉至水滾，熄火，加蓋焗約 15 分鐘至熟透，放入凍開水待冷卻，用手撕成幼絲。

2. 西芹、甘筍、甜紅椒、甜黃椒及京葱洗淨，與酸子薑分別切成約牙籤般大小的幼條，加入雞絲與預先混合的調味料拌勻，最後灑上炒熟芝麻及飛魚子即成。

必 學 重 點

■ 雞肉不宜大火煮熟，用熱水浸熟肉質才嫩滑。

■ 調味料與材料拌勻後，建議立即食用，避免材料水分滲出太多，令調味料流失，影響味道。

懷舊
千層峯

Thousand-layer pork ear terrine

6 人份量

○ 材料

大豬耳	6 隻
葱	1 棵
薑	5 片
胡椒粉	1 茶匙
魚膠粉	40 克
清水	200 毫升

○ 滷水料 A

油	1 湯匙
蒜肉片	20 克
乾葱片	30 克
葱段	30 克
川椒	半湯匙
八角	2 粒
紅椒	1 隻
花雕酒	1 湯匙（後下）

○ 滷水料 B

清水	1.3 公升
生抽	100 克
老抽	20 克
幼鹽	7 克
片糖	75 克
雞粉	10 克
蠔油	70 克
胡椒粉	1/4 茶匙
南薑	20 克
芫茜	15 克

必 學 重 點

■ 將滷水豬耳平均鋪在壽司蓆的上中部分，
　蓆的頭尾部分預留約一吋，包捲時要緊
　實，避免太多空隙，冷藏後才不鬆散。

■ 豬耳的厚身部分宜剪去，排入壽司蓆時容
　易相叠及緊壓。

做法

1. 豬耳燒毛後,洗淨,切去肥油。

2. 燒滾水約 1.5 公升,加入葱段、薑片、胡椒粉及豬耳,用慢火煮 15 分鐘,去除豬羶味,再用清水沖淨。

3. 放入竹笪,排入豬耳;另外將滷水料 A 炒香,加入滷水料 B,煮滾後放入煲內至滾起,轉慢火煲約 2 小時,熄火。

4. 魚膠粉與清水 200 毫升拌勻,待吸收水分後,用熱水座溶魚膠粉,倒入滷水內拌勻。

5. 壽司蓆平放,掃上麻油,將豬耳排好拼成方形,用力捲實並用麻繩綁緊固定,壓上重物,放於雪櫃貯存 8 小時,切片享用。

惹味小食

燒、炸、燜、煮，
小吃的多層次口感，
令人無可抗拒！

Snacks

Grilled, deep-fried, braised, stewed
Snacks come in multiple dimensions of mouthfeel
Simply irresistible

肉鬆
炸薯棗

Deep-fried sweet potato balls with
pork floss filling

清朝乾隆皇帝年老時愛吃番薯，於八十大壽時欲將番薯賞賜眾大臣品嘗，但后妃們極力反對，因當時番薯被視為民間粗俗食品，怕有損龍威。御膳房知道後想出將番薯去皮後蒸熟，加入糯米粉搓好，塑成十二生肖，再炸至金黃色，按各大臣的生肖賞賜。炸番薯口感內酥內韌，香甜可口，各人非常讚賞。

此製法流傳至民間各地，而福建省泉州地區盛產番薯，故此美味的小吃在泉州流行，家傳户曉。由於平民百姓沒有御廚的精湛手藝，簡單地搓成圓球狀，故名為炸薯棗。

可製成 12 粒

○材料

番薯	170 克
糯米粉	125 克
麵粉	30 克
熱水	100 毫升
黃糖	1.5 湯匙
黑芝麻	3 茶匙
白芝麻	7 茶匙

○餡料

花生醬	4 茶匙
豬肉鬆	12 茶匙

做法

1. 番薯洗淨，去皮、切件，隔水蒸約 20 分鐘至熟，待冷備用。

2. 黃糖放入熱水中拌溶，加入番薯搓成薯茸，待涼後，加入糯米粉及麵粉搓成粉糰。

3. 將粉糰分成 12 小粒，搓圓後壓扁，每粒加入 1/4 茶匙花生醬及 1 茶匙豬肉鬆做成餡料，包好後搓圓，在表面黏上黑白芝麻。

4. 鑊中倒入油，放入薯棗用慢火燒至油熱，炸至薯棗浮起並開始脹大時，轉大火再炸至金黃色，即可盛起，吸乾油分享用。

必學重點

- 番薯本身所含的水分各有不同，會影響搓成粉糰後的軟硬度，若太乾硬可適量地逐少加入水。

- 番薯黏上芝麻前，如出現乾裂現象，可用手指沾上少許清水掃在破裂處，使粉糰濕潤可黏上芝麻。

- 炸薯棗時必須注意油溫，最初調至低油溫，否則油溫太高令薯棗表面轉硬，不易脹大並影響口感。待薯棗受熱後浮起及脹大，方可轉大火將薯棗炸至皮脆。

五香
豬肉乾

Honey-glazed pork jerky

6 人份量

○ 材料

| 半肥瘦豬絞肉 | 300 克 |
| 錫紙 | 1 張 |

○ 調味料

百花蜜	4 茶匙	蠔油	1 茶匙
幼鹽	半茶匙	海鮮醬	1 茶匙
雞粉	半茶匙	玫瑰露酒	半茶匙
生抽	半茶匙	五香粉	半茶匙

做法

1. 豬絞肉放入大碗，加入調味料攪拌至起膠黏成一團，放入雪櫃醃 2 小時。

2. 鋪上錫紙，平均鋪上豬絞肉，蓋上保鮮紙，用擀麵棍來回擀開豬絞肉，壓成 2 毫米厚的薄肉片後，取走保鮮紙。

3. 預熱焗爐，放入焗盆用 175℃烘薄肉片每面各 8 分鐘，若焗盆有水分需倒去，將肉乾每面再烘 5 分鐘。

4. 烘焗至肉乾的油分滲出，表面呈光澤及微焦香，取出待涼食用。

必 學 重 點

- 絞肉必須攪拌至有黏性，擀薄時較平滑，烤好的肉乾才黏合成一整片。
- 絞肉不宜太粗粒，否則烤焗時難以蒸發水分，影響肉乾的效果。

果王
燒雞翼

Grilled chicken wings stuffed with
durian and cheese

6 人份量

○ 材料

雞全翼	6 隻
榴槤肉	120 克
芝士	30 克
蜜糖	3 湯匙

○ 醃料

砂糖	5 湯匙
幼鹽	1 湯匙
雞粉	半湯匙
海鮮醬	2 湯匙
蠔油	1 湯匙
芝麻醬	1 湯匙
老抽	1 湯匙
生抽	1 湯匙
薑茸	1 湯匙
乾葱茸	1 湯匙
蒜茸	1 湯匙

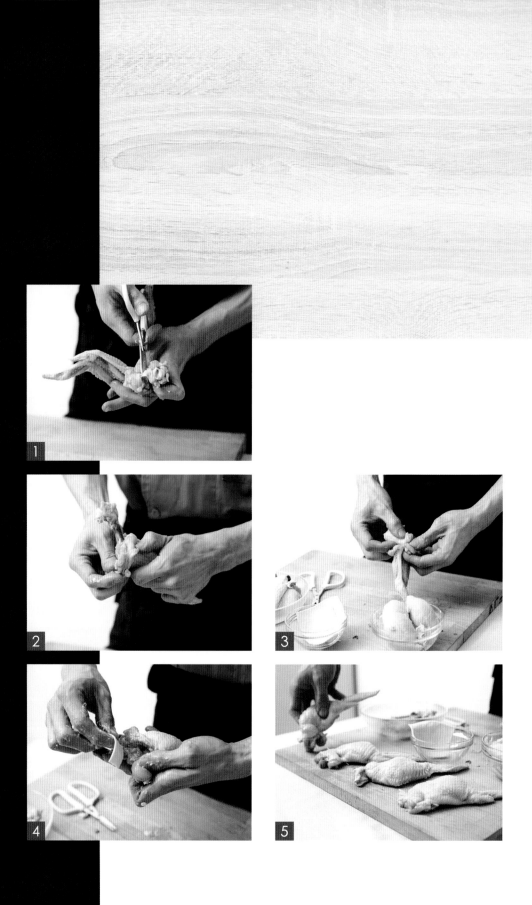

做法

1. 雞全翼洗淨，去骨。每隻雞翼釀入榴槤肉 20 克及芝士 5 克，用牙籤緊緊地封口。

2. 醃料拌勻後，放入雞翼醃 30 分鐘。

3. 焗爐預熱至 200℃，放入雞翼兩面各烘 5 分鐘，取出塗上蜜糖，再將兩面各烘約 5 分鐘即成。

必學重點

■ 去骨時小心剪破雞皮，烤製時餡料很易滲出來。

■ 釀入雞翼的榴槤肉份量不宜過多，因榴槤肉受熱後會膨脹，容易導致雞皮爆裂溢出餡料。

紅粉燒
豬頸肉

Paprika scented grilled pork cheek

6 人份量

◯ 材料		◯ 醃料		◯ 蘸料	
豬頸肉	2 件	砂糖	5 湯匙	熱水	4 克
蜜糖	2 湯匙	幼鹽	1 湯匙	金獅糖漿	30 克
玫瑰露酒	1.5 湯匙	雞粉	1 茶匙	老抽	16 克
		五香粉	1.5 茶匙	生抽	8 克
		紅椒粉	2 茶匙	青檸汁	4 克
		芫茜粉	1 茶匙	葱白	8 克（切段）
				芫茜	8 克（切段）
				指天椒	1 隻（切碎）

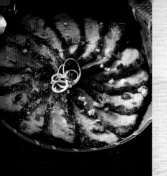

必學重點

高溫烘焗時，注意時間不可過長，否則肉質容易烘焦及乾硬。

做法

1. 豬頸肉洗淨，瀝乾水分，加入玫瑰露酒拌勻，將預先混合的醃料搽勻，醃 30 分鐘。

2. 焗爐預熱至 240℃，放入豬頸肉兩面各烘 10 分鐘，取出塗上蜜糖，放回焗爐用 200℃將兩面各烘 5 分鐘，取出備用。

3. 享用時，將豬頸肉切成薄片，淋上預先混合的蘸料即可。

Crispy honeycomb tofu

蜂 巢
脆 豆 腐

○ 材料

布包豆腐　2件

○ 滷水材料

清水	600克
八角	2粒
沙薑	10粒
丁香	1茶匙
香葉	5片
桂皮	5克
胡椒粉	半茶匙
芫茜	2棵（切段）
唐芹	40克（切段）
紅椒	1隻（切段）
薑片	10克

○ 滷水調味料

幼鹽	2湯匙
砂糖	2茶匙
雞粉	1茶匙

○ 脆漿料

麵粉	60克
澄麵	2克
生粉	2克
泡打粉	7克
清水	45克
油	45克

做法

1. 滷水材料煮滾，轉慢火加蓋煲 10 分鐘，熄火，加入滷水調味料拌至完全溶化，備用。

2. 每件豆腐切成 4 小件，隔水蒸 5 分鐘，取出放入滷水內浸泡 1 小時，盛起，瀝乾水分。

3. 燒熱油鑊，放入豆腐件炸至金黃色，瀝乾油分，待冷卻。

4. 脆漿料混和，放入豆腐件至全部黏上脆漿。

5. 燒熱油鑊，放入豆腐件炸至金黃色及皮脆即可。

必學重點

- 油鑊必須燒至高溫，炸豆腐的表面才平滑，否則容易出現氣泡。
- 豆腐放入油鑊後，待炸至硬身才推動，否則容易破損。

Deep-fried pork intestines stuffed with
minced cuttlefish

花 枝
炸 大 腸

6 人份量

○ 材料

| 豬大腸 | 1 斤 |
| 墨魚肉 | 200 克 |

○ 洗腸料

| 粗鹽 | 120 克 |
| 生粉 | 120 克 |

○ 煮大腸料

清水	200 克
白醋	3.5 湯匙
薑片	40 克
葱段	40 克
八角	3 粒
胡椒粉	1 茶匙

○ 墨魚膠醃料

鹽	1/4 茶匙
雞粉	1/4 茶匙
粟粉	2 茶匙
芫茜梗碎	1 茶匙

豬大腸上皮後，要徹底晾乾至表面乾爽，才令炸出來的外皮卜脆，或可用電風筒幫助吹乾。

○ 滷水料

清水	1.5 千克
蠔油	15 克
鹽	30 克
冰糖	20 克
雞粉	12 克
八角	3 粒
桂皮	3 克
川椒	1 湯匙
草果	3 粒
胡椒粉	1 茶匙
乾葱碎	60 克
薑片	20 克
五香粉	1 茶匙
花雕酒	1 湯匙
玫瑰露酒	1 湯匙
紅麴米	半湯匙

○ 上皮水料

清水	35 克
白醋	35 克
麥芽糖	10 克
大紅浙醋	5 克
雙蒸酒	20 克

○ 甜酸汁（拌勻）

大紅浙醋	1 湯匙
黃糖	1 湯匙
蘇梅醬	1 湯匙
茄汁	1 湯匙

做法

① 清洗大腸

1. 豬大腸翻出內壁，用洗腸料揉搓，將黏液污物徹底清除，再用清水洗淨（重覆此步驟三次），翻回內壁，用清水洗淨。

2. 煲滾煮大腸料，放入豬大腸至滾，轉慢火煲 80 分鐘，取出，用清水沖淨，瀝乾水分。

② 墨魚膠

1. 墨魚肉洗淨，抹乾水分，切小件，放入攪肉機攪碎。

2. 墨魚漿放入碗內，加入醃料順一方向攪拌至起膠即可。

③ 滷釀大腸

1. 煲滾滷水料，放入豬大腸用慢火煲 30 分鐘，取出，用熱水沖淨表面，掛起瀝乾水分。

2. 上皮水料用小火煮熱，至麥芽糖溶解。

3. 豬大腸表面塗勻上皮水，風乾備用。

4. 將墨魚膠釀入豬大腸內。鑊中燒油至 7 成熱，下豬大腸炸至皮脆，享用時蘸甜酸汁品嘗。

雞卷，是福建閩南地方富有特色的民間傳統美食。每到新年、喜宴、祭祀之時，總會炮製雞卷，成為餐桌上的必然食品。

雖名為雞卷，但材料卻沒有雞肉成份，究竟菜名如何得來？民間有說從前辦酒席時，廚師會將多餘的剩菜、豬肉，用豬網油捲起來炸透，並取名「多卷」，閩南語的「多」跟「雞」的發音類似，於是演變成「雞卷」。

6 人份量

○ 材料

腐皮	1 張
五花腩	200 克
馬蹄肉	60 克
洋蔥	40 克
乾蔥頭	40 克
蔥白	40 克
番薯粉	5 湯匙
麵粉	2.5 湯匙
雞蛋	1 個

○ 醃料

生抽	1 湯匙
砂糖	半茶匙
花雕酒	2 茶匙
幼鹽	半茶匙
五香粉	半茶匙

○ 麵粉糊 （拌勻）

麵粉	1 湯匙
清水	1.5 湯匙

閩南
五香雞卷

Five-spice pork roll in Minnan style

必學重點

蒸肉卷的時間要控制得宜，時間不能太長，否則包肉卷的腐皮被蒸至破爛，影響外表及口感。

做法

1. 五花腩洗淨，與馬蹄肉、洋葱、乾葱頭及葱白分別切成小粒（約黃豆般大小），加入醃料拌勻醃約 15 分鐘，再下番薯粉、麵粉及雞蛋揉搓至起膠，即成肉餡。

2. 腐皮剪成 4 小份，用熱水塗勻腐皮兩面使其略軟身。

3. 將肉餡分成 4 小份，每份放上腐皮內捲實，用麵粉糊封口。

4. 預先在蒸架掃上一層油，放上肉卷隔水蒸 25 分鐘，取出待冷。

5. 將肉卷切成約半吋厚片，放入滾油炸至金黃色及皮脆，可趁熱享用。

鮑汁
焖鳳爪

Chicken feet braised in abalone sauce

◯ 材料

雞腳 12 隻

◯ 鮑魚汁料

清水	1.4 公升	八角	3 粒
蠔油	60 克	蝦米	30 克
砂糖	20 克	大地魚乾	30 克
老抽	120 克	薑片	20 克
花雕酒	1 湯匙	葱段	20 克
五香粉	1 茶匙	乾葱肉	20 克（切片）
香葉	4 片		

必 學 重 點

鮑魚汁的配方原是用作扣製乾鮑魚之用，用後再翻熱，待涼後可貯存雪櫃留作其他用途，如鮑汁燜伊麵、鮑汁滷鳳爪等。

做法

① 鮑魚汁

1. 燒熱油,放入薑片、乾葱片、葱段、蝦米及大地魚乾,炸香至金黃色,轉放煲內。

2. 與其他鮑汁料混和,用慢火煲滾,熄火備用。

② 燜鳳爪

1. 雞腳洗淨後,剪去腳甲,放入滾水煮約 2 分鐘,盛起,浸泡清水至冷卻,盛起備用。

2. 鮑魚汁煮滾後,放入竹笪後排入雞腳,待翻滾,轉慢火煲約45 分鐘即成。

蠔 仔 芋 圓

Taro balls in baby oyster soup

蠔仔芋圓是福建閩南地區的傳統小吃，據説芋圓早
於幾百年前已出現，相傳在石獅市的海邊，漁民出
海捕魚難以攜帶食物，有人靈機一觸，將芋頭去皮
及搗碎，拌入番薯粉並搓成圓條狀，蒸熟後晶瑩剔
透，口感有嚼勁，味道很好，冷吃熱吃皆宜，作為
乾糧最合適，於是成為漁民出海必備的獨特美食，
並名為芋圓。

在眾多烹煮方法中，以蠔仔搭配芋圓煮成湯最廣為
流傳，因蠔仔的鮮甜味與芋圓相當匹配。

6 人份量

◯芋圓材料

| 荔芋 | 200 克 |
| 番薯粉 | 80 克 |

◯芋圓調味料

熱水	2 湯匙
鹽	1/4 茶匙
雞粉	1/4 茶匙

◯湯料頭

油	1.5 湯匙
薑片	25 克
乾葱片	25 克
南薑片	20 克

◯湯調味料

| 鹽 | 1 茶匙 |
| 雞粉 | 半茶匙 |

◯湯材料 A

清水	600 克
荔芋茸	100 克（刨茸）
娃娃菜	100 克
芋圓	約 18 粒

◯湯材料 B

蠔仔	200 克
芫茜	1 棵（切段）
葱段	1 棵（切段）
大蒜	1 棵（切段）

做法

①芋圓

1. 荔芋 200 克刨成茸,加入番薯粉 80 克與預先混合的調味料揉成粉糰,
   用手唧成魚蛋狀,每粒 15 克,約做 18 粒。
2. 蒸籠或蒸架先掃上油,放上芋圓隔水蒸 15 分鐘,即成芋圓。

②蠔仔芋茸湯

1. 燒熱油鑊,放入湯料頭爆香,下湯材料 A 至滾,拌入湯調味料,轉慢
   火煮 5 分鐘。
2. 最後,加入預先洗淨的湯材料 B 至滾,再煮約 2 分鐘即成。

必 學 重 點

- 蠔仔烹煮時間不要太久,剛熟即可,否則失去鮮味及口感。
- 搓揉後的芋圓宜蒸熟,若下滾水燙熟,番薯粉會黏着鍋底。
- 用手唧芋圓前,建議在雙手塗上油,以免芋茸黏着手不方便。

懷舊燒味

懷舊燒味……令人情有獨鍾，
脆皮燒鵝、富貴雞，
換個角度在家炮製，吃出舊有情懷！

Barbecue meat

Everyone has their fond memories with barbecue meat
Crispy roast goose, beggar's chicken
Make them at home and have a taste of the good old days

富 貴 雞

Beggar's chicken
(roast chicken in salt crust)

video 教學

相傳清初，在江蘇常熟有一位名為「叫化」的乞丐，偶然捕獲雞隻，卻苦無炊具，他忽發奇想，將雞裹滿泥巴並投入火堆中烤熟，拍去泥土後雞毛隨泥巴脫落，由於食用時香味四溢而被人發現，隨後相繼仿做，以乞丐雞或叫化雞命名。

經過多番改良，現今以「富貴雞」之美名出現。選用肉質肥嫩之小母雞，餡料以增添鮮味的配料為主，而且多元化，炒香後填入雞腹，最後以幼鹽及蛋白代替泥巴包裹雞身，乾淨、美觀之餘，也縮短烤製時間，令雞肉保持嫩滑。

○ 材料

光雞	1 隻（約 2 斤）
乾荷葉	1 張（大）
錫紙	2 張（每張 30 厘米 x 35 厘米）

○ 醃雞味料

老抽	1 湯匙
幼鹽	3 茶匙
砂糖	2 茶匙
雞粉	1 茶匙
五香粉	1 茶匙

○ 醃雞材料

鮮沙薑茸	4 湯匙
乾葱茸	4 湯匙
薑茸	4 湯匙
生抽	半湯匙
花雕酒	4 湯匙

○ 餡味料

清水	100 克
老抽	3 茶匙
生抽	1 茶匙
蠔油	1 茶匙
幼鹽	少許

○ 餡料

油	1 湯匙
肥豬肉	50 克（切丁）
洋葱	30 克（切粒）
八角	1 粒
蒜肉	15 克
乾冬菇	3 朵（切條）
雞髀菇	50 克（切條）
細圓栗子	6 粒
麻油	半茶匙

○ 包雞泥料

雞蛋白	120 克
幼鹽	650 克

○ 埋獻

生粉	1 湯匙
清水	2 湯匙

- 用醃雞味料醃四十分鐘後,先用清水沖去味料,否則雞肉較鹹。
- 建議用打蛋器,將包雞泥料打至蛋白企身發脹為佳。
- 以蛋白料代替傳統的泥料,焗製時間短,雞肉腍滑。

做法

1. 乾荷葉用熱水浸約 30 分鐘,備用。

2. 光雞掏去肥膏、內臟,洗淨,用刀背將雞頸骨敲斷,雞背敲至稍平,轉身
   雞胸向上,用力將雞胸稍壓平,在雞胸與雞髀間的皮剝開,令雞髀鋪平。

3. 用醃雞味料的老抽搽勻全身雞皮,再用其他味料搽勻雞身,醃 40 分鐘,
   用清水沖去味料,瀝乾水分,放入油鑊煎至雞皮金黃色。

4. 雞平放盤內,胸向上,將醃雞材料拌勻,塗勻雞身醃 30 分鐘。

5. 餡料炒香後,加入餡味料加蓋慢火煮 10 分鐘,埋獻拌勻,釀入雞腔內,
   用荷葉包好,再包上錫紙。

6. 包雞泥料拌勻後,塗少許於烘盆上,放入焗爐用 200℃烘約 3 分鐘至定形,
   放上包好的雞,均勻地在錫紙上塗抹包雞泥料,緊緊地將雞密封。

7. 預熱焗爐,用 200℃焗 70 分鐘,敲碎蛋白料,即可品嘗嫩滑雞肉。

翠盞
金錢雞

Grilled chicken liver and
pork medallions

6 人份量

◯材料

肥頭肉	半斤
肉眼	半斤
大雞肝	10 副
薑	15 片
酸子薑	12 片（伴食）
西生菜	12 片（伴食）

◯醃料

砂糖	5 湯匙
幼鹽	1 湯匙
雞粉	半湯匙
五香粉	1/8 茶匙
乾葱茸	2 粒
蒜茸	3 粒
生抽	半湯匙
老抽	半湯匙
南乳	半小塊
柱候醬	半湯匙
芝麻醬	半湯匙
玫瑰露酒	1 湯匙
雞蛋	1 個

這是來自順德的名菜,在戰後貧困的年代,燒臘店的師傅物盡其用,將下欄物資收集下來,例如宰雞後棄掉的雞肝;肉檔餘下的肥豬肉及乾巴巴沒半點脂肪的瘦肉,重新研製成三層式烤肉串,燒烤後尤如金錢狀,而且甘香美味,故稱為「金錢雞」。

古法製作時,肥豬肉及瘦肉修切得較厚,吃了感膩滯。現時將肥肉及瘦肉的厚度切得更薄,配合生菜、酸子薑伴吃,口感豐富,而且又可化解肥膩感。

預備做法

① 肥頭肉

1. 肥頭肉切成邊長 4 厘米 x 4 厘米的正方形，放入滾水煮 10 分鐘，用
   清水沖洗至冷卻，切成一片片約 0.3 厘米厚的薄片，瀝乾水分，拌
   入糖 5 湯匙醃 10 小時或以上。

2. 烤製前，須提早最少 20 分鐘加入玫瑰露酒 1 湯匙拌勻，備用。

② 肉眼

肉眼放於冰櫃，冷凍至稍硬身，切成如肥頭肉般大小的薄片，加入生粉
2 湯匙、清水 2 湯匙拌勻醃 30 分鐘，備用。

③ 雞肝

雞肝用花雕酒 2 湯匙、薑碎 2 湯匙拌勻醃 30 分鐘，備用。

綜合做法

1. 醃料混和後，加入肥頭肉、肉眼及雞肝拌勻醃 20 分鐘，用叉燒針或竹籤依次串上薑片、肉眼、雞肝、肥頭肉、薑片，即成金錢雞串。

2. 焗爐預熱至高溫後，放入金錢雞串用 250℃焗約 5 分鐘至表面略見黑糖焦色，轉 125℃焗 20 分鐘，取出塗抹蜜糖。

3. 金錢雞串再用 250℃烘至表面蜜糖沸騰（約 3 分鐘），轉 100℃焗 15 分鐘，取出，再塗上蜜糖即可。

4. 享用時，西生菜修剪成小圓碗形，放上酸子薑片，排入金錢雞即可。

必 學 重 點

■ 肥頭肉是豬耳後脖子皮下，第一層較結實的肥油，宜切出最上一層肥肉烘焗，肉質較爽。

■ 肥頭肉用糖醃十小時，烘後的肥頭肉更爽口。

■ 叉燒針串上材料時，一層接一層即可，別壓得太緊密，否則材料較難熟透。

薑 母 鴨

Ginger duck soup

此菜源自福建泉州，是漢族傳統的名菜，調補氣血，配合鴨肉更有滋陰降火之效。輾轉流傳至台灣，經多番改良後，解決了藥味過濃，以及酒含量過多帶來的苦澀問題，成為了大受歡迎的名菜。

8 人份量

○ 材料

番鴨	半隻（約 2 斤半）

○ 湯料頭

黑麻油	3 湯匙
薑片	45 克
八角	2 粒
米酒	80 克（後下）

○ 湯材料

清水	500 克
薑片	40 克
當歸	2 克
川芎	4 克
參根	3 克
北芪	6 克
肉桂	3 克
圓肉	2 粒
杞子	6 克（後下）

○ 湯調味料

頭抽	1.5 湯匙
黑糯米酒	50 克

○ 蘸汁料頭

黑麻油	1 湯匙
薑粒	1 湯匙
乾葱粒	1 湯匙
蒜粒	1 湯匙
花雕酒	1 湯匙
黑糯米酒	1 湯匙

○ 蘸汁調味料

清水	5 湯匙
生抽	2 湯匙
老抽	半湯匙
鹽	1/4 茶匙
雞粉	1/4 茶匙
黃糖	2 湯匙

做法

1. 番鴨去除內臟及尾部肥油，洗淨，斬件。

2. 燒熱煎鑊，放入鴨件（皮向下）用慢火煎至鴨皮呈金黃色，略炒香後盛起。

3. 燒熱油鑊，放入湯料頭爆香，下鴨件及米酒炒香，加入湯材料至滾，轉慢火，加蓋煮 1 小時，下湯調味料拌勻即成。

4. 燒熱油鑊，放入蘸汁料頭爆香，拌入蘸汁調味料煮滾，伴鴨件享用。

必學重點

■ 鴨湯內酒的含量較多，不宜用鹽調味，否則帶苦澀味。

■ 煮湯時，杞子宜在最後半小時才放入，否則煮太久令味道變酸。

深 井 脆 皮 燒 鵝

深井以燒鵝美食馳名半個世紀，源於早年務農的村民
以養鵝為生，並將鵝隻燒烤出售，供應鄰近的工人及
假日的遊客享用，漸受歡迎。

傳統做法以炭火燒烤，香味獨特，但因燒炭火的牌照
問題，現今改用煤氣爐或電烤爐烤製，而深井燒鵝主
要的特點是控制火候及出爐的時間，趁熱享用，鵝肉
香氣飄散，皮脆肉嫩，汁多味美。

Crispy roast goose in
Shum Tseng style

○ 材料

黑棕鵝　　1 隻（約 5 斤）

○ 醃料 A

砂糖	4 湯匙
幼鹽	1 湯匙
雞粉	1 茶匙
芝麻醬	半湯匙
海鮮醬	半湯匙
蠔油	1 湯匙
花雕酒	1 湯匙

○ 上皮料

白醋	120 克
麥芽糖	48 克
大紅浙醋	6 克

○ 醃料 B

八角	3 粒
薑	3 片
芫茜	1 棵
乾葱頭	3 粒

○ 蘸料

燒味汁	100 克（做法參考 p.10）
酸梅醬	3 湯匙

必 學 重 點

- 燒鵝烘熟後，內腔的鵝肉汁滾燙，小心拔出鵝尾針時被鵝肉汁燙傷。
- 燒鵝炸脆後，建議掛起待十分鐘，定型後才斬件，肉汁不會流失。

做法

1. 黑棕鵝去肥油、內臟及喉管，洗淨，瀝乾水分。

2. 醃料 A 混和後，倒入鵝腔內搽勻，再放入醃料 B，用鵝尾針把開口位穿好。

3. 燒熱水，放入黑棕鵝燙勻全身，至鵝皮收緊變色，取出即用清水沖洗至鵝皮涼爽，勾起瀝乾水分。於鵝皮塗勻上皮料，風乾備用。

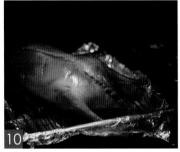

4. 焗爐預熱至 150℃，黑棕鵝胸部向上平放在焗盆，烘約 50 分鐘至表皮呈
   淺紅色，蓋上錫紙轉 100℃烘 30 分鐘至熟透，取出掛起及拔出鵝尾針，
   令內腔的水分滴乾。

5. 燒熱油約 200℃，將燒鵝炸至皮脆，斬件上碟，蘸燒味汁及酸梅醬享用。

甜醋雞

Marinated chicken in
sweet vinegar sauce

這是香港另類懷舊菜式，早期的酒樓以豬腳薑作為小吃，廚師
將餘下的甜醋汁加以利用，配以豉油雞烹煮，將甜醋汁的獨特
香味滲透雞肉，吃起來特別醒胃可口。

6 人份量

○材料

光雞	半隻（約 650 克）
薑	3 片
葱	15 克（切段）

○甜醋汁料

八珍甜醋	1.5 公斤
老薑或子薑	450 克
豬手件	450 克（約 6 件）

○滷水料

清水	900 克	玫瑰露酒	40 克
生抽	900 克	五香粉	2 湯匙
老抽	16 克	沙薑粒	25 克
黃糖	400 克	八角	4 粒

甜醋汁做法

1. 薑刮去外皮，用刀略拍至裂開。
2. 豬手去掉毛及蹄甲，放入滾水內，下薑片及葱段用慢火煮 30 分鐘，去除血水，取出沖淨，瀝乾備用。
3. 燒熱瓦煲，下少許油爆香老薑或子薑，放入豬手及甜醋煮滾，加蓋用慢火煲 90 分鐘即可。

綜合做法

1. 光雞去掉毛及內臟，洗淨備用。
2. 滷水料用慢火煮滾，攪拌至黃糖完全溶化，放入光雞至滾，以慢火再煮 5 分鐘，熄火，加蓋焗 30 分鐘至熟，盛起。
3. 滷水雞斬件後，放入鍋內，倒入甜醋汁煮滾即成。

狗仔鴨

Doggie duck pot

此菜是肇慶的名菜，烹調方法與煮狗肉做法相似
而得名。由於狗肉絕非大部分人能接受，但這種
味道確實吸引很多人，故改以鴨肉取替，味道卻
很匹配，後來廣泛流傳，更被稱為狗仔鴨。

6 人份量

○ 材料

米鴨	半隻（約 1 公斤）
蓮藕	300 克

○ 滷水料頭

油	1 湯匙
薑片	20 克
乾葱頭	30 克（切片）
蒜肉	10 克（切片）
八角	3 粒
丁香	半湯匙
白豆蔻	10 粒
花雕酒	2 湯匙

○ 滷水材料

清水	900 克
白腐乳	50 克
回味醬	4 湯匙
	（做法參考 p.12）
陳皮	半個
沙薑粒	8 粒
幼鹽	半茶匙
雞粉	半茶匙
生抽	1 湯匙
老抽	1 湯匙

○ 生粉獻

清水	2 湯匙
生粉	1 湯匙

做法

1. 米鴨去掉肥油及內臟，洗淨，斬件備用。

2. 蓮藕去皮，洗淨，切成圓片狀。

3. 鴨件放於煎鍋內，鴨皮向下用慢火煎至表皮金黃色，鴨肉稍煎至少許焦糖色，盛起備用。

4. 燒熱油，下滷水料頭爆香，加入滷水材料煮滾，下鴨件及蓮藕煮滾，轉慢火加蓋燜 50 分鐘，最後下生粉獻即成。

必 學 重 點

斬鴨件時建議去掉鴨尾部位，因羶味太重，影響食味。

芋 泥 缽 仔 鵝

這是由歷史悠久的傳統菜式「芋頭燜鵝」演變而成，曾於六、七十年代的香港較流行。現時的做法將芋頭製成芋泥，散發更濃郁的芋香味，鵝肉以滷水浸至入味，再配合芋泥烹煮，令兩種香味互相輝映，使人回味無窮。

Braised goose with mashed taro in earthenware bowl

○ 材料

黑棕鵝	1 隻（約 5 斤半）
荔芋件	24 件
葱、芫茜	各 1 棵

○ 醃鵝腔料

砂糖	3 湯匙
幼鹽	1 湯匙
雞粉	半湯匙
五香粉	半湯匙
八角	2 粒
薑	3 片
葱	1 棵

○ 滷水料

清水	3.6 公斤
幼鹽	5 湯匙
雞粉	1.5 湯匙
老抽	9 湯匙
八角	6 粒
沙薑	12 片
香葉	9 片
薑	30 克
花雕酒	3 湯匙

○ 鵝骨湯材料

鵝頭	1 個
鵝頸	1 條
鵝翼	1 對
鵝腳	1 對
薑	3 片

○ 芋茸料

荔芋肉	350 克
鵝骨湯	350 克

必 學 重 點

■ 煮缽仔鵝時不可用大火，否則芋泥汁容易煮焦及黏底。

■ 煲完鵝骨湯的材料可放入滷水料煲至入味，再與鵝肉同時烹煮。

■ 用鵝尾針在鵝腔位穿上一半，使熱力進入鵝腔內，令鵝肉容易熟透，並防止鵝籠汁流走。

◯ 炸料頭

紅葱頭	4 粒
薑	6 片
蒜頭	6 粒

◯ 芋泥汁

熟鵝油	4 湯匙（取鵝油膏煮成）
紅葱頭	50 克
蒜頭	30 克
薑片	30 克
花雕酒	2 湯匙
鵝籠汁	120 克（做法見步驟 3）
鵝骨湯	200 克
芋茸	700 克
柱候醬	1 湯匙
芝麻醬	1 湯匙
蠔油	1 湯匙

預備做法

① 滷水汁

滷水料煮滾後，轉慢火煮 15 分鐘，熄火，待冷。

② 芋茸

1. 將鵝骨湯材料洗淨，放入 1.2 公升的清水煮滾，轉慢火煲 90 分鐘，隔渣後，餘下的湯撇去油即成鵝骨湯（分成 350 克及 200 克使用）。

2. 荔芋肉 350 克切件，隔水蒸 20 分鐘，待冷後，加入鵝骨湯 350 克攪拌成芋茸。

③ 荔芋件及炸料頭

1. 荔芋件切成 24 件，每件約 1 吋 x 2 吋（厚度約 6 毫米），燒熱油，放入芋件炸至金黃色取出。

2. 燒熱油，放入炸料頭炸至金黃色，取出備用。

④ 芋泥汁

1. 鵝油膏放於鍋內，用慢火煮至油塊收乾成金黃色油渣，即成熟鵝油（見圖 1 及 2）。

2. 燒熱鑊，放入熟鵝油燒熱，下紅葱頭、蒜頭及薑片炒至金黃色，灒花雕酒，加入其他材料用慢火煮滾，熄火備用。

做法

1. 光鵝取出鵝油膏（用以製成熟鵝油），去除內臟，斬去腳、翼、頸、頭（用於鵝骨湯），洗淨備用。

2. 將醃鵝腔料放入鵝腔內，搽勻及醃 30 分鐘，用鵝尾針在開腔位穿上一半，留一半開口。

3. 預熱蒸鑊，將鵝身平放（胸向上），隔水蒸約 1 小時，熟透後取出，待稍降溫後，倒出「鵝籠汁」，隔渣備用。

4. 原隻鵝斬成 4 大件，放入滷水內醃 5 小時或以上，切成鵝片。

5. 將葱、芫茜及炸料頭放缽仔底，排上炸芋件及鵝片，倒入芋泥汁，用慢火煮滾即成。

串串
麻辣燙

Spicy hot pot in
Sichuan style

此菜起源自四川省近長江流域地區，至今已有幾百年歷史。當時，生活於長江的船運工人，為了煮食便利，在江邊隨地疊起石塊及架起瓦罐，倒入江水、調味料，用乾柴生火燒滾火鍋湯，放入用竹籤串成的材料燙熟，風味獨特。

現時，火鍋的吃法變得多元化，湯底選擇多，如豬骨湯、雞湯、牛骨湯等，各種風味能滿足不同的需求。

6 人份量

○ 材料

豬骨	600 克
薑片	20 克
清水	1.8 公斤

○ 湯材料

豬骨湯	900 克
紅椒粉	1 湯匙
五香粉	1.5 茶匙

○ 湯料頭

牛油	1.5 湯匙
油	1 湯匙
薑片	15 克
葱段	15 克
乾辣椒	3 隻
蒜片	15 克
乾葱片	30 克
豆瓣醬	2 湯匙
回鍋醬	3 湯匙（做法參考 p.11）
花椒油	1 湯匙
辣椒油	1 湯匙

○ 湯調味料

鹽	1 茶匙
雞粉	半茶匙
蠔油	1 茶匙

做法

1. 豬骨飛水後，放入鍋內加薑片及清水，用慢火煲約 3 小時至餘下豬骨湯約 900 克，熄火備用。
2. 湯料頭爆香，加入湯材料煮滾，下調味料拌勻，即成麻辣湯底。
3. 火鍋食材用竹籤穿成一串串，放進湯底內燙熟即可享用。

必 學 重 點

- 麻辣湯的辣度可隨個人口味調較，增減回鍋醬、乾辣椒及辣椒油的份量即可。
- 串燙的食材可選用較易吸收麻辣湯味道的材料，如魚丸類、小塊肉類、豆卜及蔬菜菌類等。

燒味搭上了海鮮，出奇地配合，
回味的醬汁精華，
令鮮味提升不少。

Seafood

Seafood and barbecue meat may not make a conventional combo
But they taste unexpectedly good together
Homemade sauces and seafood juices complement each other perfectly

○ 材料

響螺肉	250 克
蟲草花	20 克
金華火腿	20 克
薑片	20 克
杞子	12 粒
清水	1.2 公升

○ 包紮材料

椰菜葉	6 片
韭菜	6 條

○ 調味料

鹽	半茶匙
雞粉	1/4 茶匙
胡椒粉	1/8 茶匙

蟲草花
響螺球

Cabbage rolls filled with
conch and cordyceps flowers in soup

做法

1. 響螺肉去除內臟，用粟粉 2 茶匙及鹽 1 茶匙洗擦乾淨，以清水洗淨。

2. 金華火腿洗淨，與響螺肉一起飛水。

3. 燒滾水 1.2 公升，放入金華火腿、響螺肉、蟲草花及薑片，用慢火煲 3 小時， 最後加入杞子及調味料再煲 30 分鐘，至餘下湯水 600 克，熄火，隔出湯料分成 6 等份。

4. 椰菜葉及韭菜用滾水稍燙至軟身，取椰菜葉 1 片，放上湯料 1 份，用韭菜綁緊成蔬菜球。

5. 將蔬菜球放回湯內，用慢火煮 5 分鐘即成。

川汁
雞煲蟹

Chicken and crab hot pot
in Sichuan style

○材料

光雞	1隻（約2斤4兩）
肉蟹	1斤

○醃料

生抽	1湯匙
老抽	半湯匙
花雕酒	1湯匙
砂糖	1湯匙
雞粉	半湯匙
五香粉	1茶匙

○川汁料頭

油	2湯匙	紅椒	2隻
薑片	30克	八角	2粒
乾葱片	40克	川椒粒	2湯匙
蒜片	40克	沙薑片	10粒
葱段	1棵	花雕酒	2湯匙（後下）
乾辣椒	2湯匙		

必 學 重 點

川椒粒爆香後，用茶包袋裝好，以免在烹煮時川椒碎粒黏在其他材料上，影響口感。

○ 川汁調味料

清水	300 毫升	豆瓣醬	1 茶匙
老干媽油辣椒	2 湯匙	唐芹段	1 棵（後下）
蠔油	1 湯匙	芫茜段	1 棵（後下）
海鮮醬	1 湯匙		

做法

1. 光雞去除內臟、肥油後，洗淨，斬件，加入醃料拌勻醃 15 分鐘。
2. 雞件取出，瀝乾，放入煎鍋用慢火煎至金黃色備用。
3. 肉蟹用刀斬去蟹箝，反轉蟹身，蟹肚朝天，去除奄蓋，揭開蟹蓋，拆掉鰓部份，斬件及洗淨，用刀將蟹箝拍至略破裂。
4. 燒熱砂鍋，放入川汁料頭爆香，下雞件炒勻，瀸酒，倒入川汁調味料，待水滾後轉慢火，加蓋煮 10 分鐘，最後加入蟹件炒勻，下唐芹及芫茜，加蓋再煮 3 分鐘即成。

鴛鴦醬
焗扇貝

Grilled scallops in duo sauces

6 人份量

◯ 材料

大扇貝　　6 隻

◯ 調味料

回鍋醬　　6 湯匙（做法參考 p.11）
回味醬　　6 湯匙（做法參考 p.12）
蒜茸　　　3 茶匙

做法

1. 扇貝取肉及群邊（外殼保留），用粟粉搓揉扇貝肉，洗淨。
2. 將扇貝肉及群邊放入窩狀的半殼內，調味料混合後塗勻扇貝肉。
3. 焗爐預熱至 240℃，放入扇貝烘約 4 分鐘，趁熱享用。

必 學 重 點

- 選購扇貝時，必須選閉殼的，如貝殼張開或貝肉容易
  脫落，表示不新鮮。
- 烘焗扇貝的時間不可太久，否則扇貝肉收縮變韌。

○材料

銀鱈魚	450 克

○醃料

清水	240 克	雞粉	1 茶匙
鹽	2 茶匙	花雕酒	1 湯匙
砂糖	1 茶匙	蒜茸	30 克

○黑蒜汁料頭

油	2 茶匙
薑粒	5 克
葱段	10 克
蒜片	10 克
米酒	1 茶匙（後下）

○黑蒜汁調味料

黑蒜	15 克（切幼粒）
燒味汁	3 湯匙（做法參考 p.10）
清水	3 湯匙

○獻汁

粟粉	1/4 茶匙
清水	半茶匙

做法

1. 銀鱈魚洗淨，切成 6 件，去骨，放入醃料醃約 25 分鐘，取出瀝乾水分，備用。

2. 燒熱鑊，下少許油爆香黑蒜汁料頭，灒酒，加入黑蒜汁調味料用慢火煮滾，埋獻拌勻，熄火備用。

3. 魚塊與粟粉 2 湯匙拌勻，燒熱油鑊，用慢火將魚件每面煎至金黃色，上碟，淋上黑蒜汁即成。

必學重點

魚件必須除去骨刺，否則煎魚時魚肉收縮，令魚骨突出，魚肉不能貼着鑊面煎至熟透。

黑蒜汁
銀鱈魚

Pan-fried black cod in
black garlic sauce

玫瑰
煙鯧魚

Smoked rose-scented pomfret

○ 材料

鯇魚	1 條（約 14 兩）

○ 調味料

清水	300 克
法國乾玫瑰	4 湯匙
紅麴米	1 湯匙
麥芽糖	50 克
砂糖	30 克
生抽	30 克
老抽	10 克
雞粉	5 克
幼鹽	13 克

○ 雜菜汁材料

甘筍	160 克
洋葱	80 克
西芹	60 克
芫茜	10 克
蒜肉	10 克
乾葱肉	10 克
薑肉	10 克
南薑	20 克
玫瑰露酒	4 茶匙

○ 煙燻料

法國乾玫瑰	5 朵
麵粉	1 湯匙
砂糖	1 湯匙

預備做法

1. 調味料做法：清水、乾玫瑰及紅麴米煮滾，轉慢火煮 5 分鐘，加入其他調味料攪拌至完全溶化，熄火待冷，備用。
2. 雜菜汁做法：雜菜汁材料全部切碎，放入攪拌機，加入上述的調味料攪打成汁。

做法

1. 魚不用在肚腹位置切開，斜刀切件，去內臟，洗淨，抹乾，下雜菜汁醃 10 小時。
2. 焗爐預熱至 240℃，放入鎗魚每面各烘約 6 分鐘至熟透。
3. 煙燻料拌勻，用錫紙盛好，平放於鍋中間，排入蒸架，將鎗魚件排在架上，開火至滲出煙時，加蓋調慢火焗 3 分鐘即成。

必 學 重 點

煙燻時間不宜太長，因煙燻時有一定溫度，否則令魚肉在焗後更鬆散。

蜑家海鮮
泡飯珠

Fisherman's rice balls in seafood soup

這是蜑家人日常的食法，今天稍加改良，
卻仍保留了飯珠的做法。

○材料

石斑魚	1 條（600 克）
海蝦	6 隻（約 300 克）
白飯	240 克
菠菜葉	12 片
蒜茸	1 茶匙
芫茜碎	10 克
唐芹碎	30 克
葱花	20 克

○醃料

醃味白滷水汁　250 克（做法參考 p.13）

○菠菜調味料

幼鹽	適量
雞粉	適量

○湯材料

蝦殼	6 隻
魚骨	1 條（魚的剩骨）
薑片	20 克
小茴香	1 茶匙
胡椒粉	1/4 茶匙
清水	1.8 公斤

○湯調味料

幼鹽	半茶匙
雞粉	1/4 茶匙

必 學 重 點

- 如使用隔夜剩飯，必須蒸熟後才進行搗爛的步驟。
- 搓飯珠時，應先在雙手塗上少許油，白飯不會黏着手，搓揉時
更容易處理。

做法

1. 石斑魚起肉，切成小塊，魚骨留用。

2. 海蝦去殼、去腸，蝦殼留用。

3. 燒熱鑊，放入油 1 茶匙，加入菠菜及蒜茸略炒，灑入調味料拌炒，盛起，待冷後切碎。

4. 白飯用石椿搗爛成粉糰狀，加入菠菜粒揉搓均勻，分成 12 小份，搓成小圓球飯珠。

5. 魚肉及蝦仁放入醃料醃 20 分鐘，用清水略沖走表面醃味水，加粟粉 2 湯匙拌勻。

6. 湯材料煮滾後，轉慢火煲約 1 小時至餘下湯 900 克，隔渣，加入湯調味料拌勻。

7. 飯珠放入湯內，用慢火煮 5 分鐘，加入魚肉及蝦仁煮滾，煮約 2 分鐘，熄火， 最後加入芫茜碎、唐芹碎及葱花即可。

回鍋醬
蒸蟶子

Steamed razor clams with
Twice-cooked sauce

6 人份量

○ 材料

蟶子	6 隻（約 500 克）	
回鍋醬	6 茶匙（做法參考 p.11）	
燒味汁	3 茶匙（做法參考 p.10）	
葱花	6 茶匙	

做法

1. 蟶子開邊後，去內臟，沖走沙泥，洗淨，瀝乾水分。

2. 排好蟶子，每隻塗上回鍋醬 1 茶匙，醃 3 分鐘。

3. 燒滾水，放入蟶子隔水蒸 3 分鐘，取出，灑上葱花，最後淋上燒味汁即可品嘗。

必 學 重 點

蟶子未烹煮前，先放在淡鹽水內待二至三小時，令蟶子吐淨泥沙，進食時口感較佳。

甘筍汁
紫蘇象拔蚌

Geoduck in perilla carrot puree

5人份量

○材料

象拔蚌	600 克（約 5 隻）
唐芹	45 克（切粒）
芫茜	15 克（切粒）
葱花	15 克

○湯材料

清水	900 克
甘筍	200 克
薑片	60 克
紫蘇葉	7 片
香葉	5 片

○調味料

幼鹽	1.5 茶匙
生抽	1 茶匙
雞粉	1 茶匙

必學重點

煮象拔蚌以剛熟為佳，如烹煮時間過長令肉質變韌。

做法

1. 用小刀剝開象拔蚌殼，去掉啡黑色內臟，洗淨備用。

2. 將清水 900 克及甘筍用攪拌機打成甘筍汁，放入鍋內煮滾，下薑片、紫蘇葉及香葉，加蓋用慢火煮 5 分鐘，灑入調味料拌勻。

3. 象拔蚌放入湯內，慢火略煮 2 分鐘，最後下唐芹、芫茜及葱花即成。

歡聚宴客

巧手弄製的宴客燒味，
單看做法已令人嘖嘖稱讚，
還有那令人垂涎的味道！

Banquet cuisines

Sumptuous barbecues fit for a gourmet banquet
Surprise your guests with the awe-inspiring steps
Enchant your guests with a feast on their palate

○材料

乳豬	1 隻 2.4 公斤（約 4 斤）

○醃料

砂糖	3 湯匙
幼鹽	2 湯匙
雞粉	半湯匙
五香粉	半茶匙

○上皮料

A 類上皮水　4 湯匙
（做法參考 p.13）

○糯米飯料

糯米	900 克
清水	1.2 公升
紫番薯	300 克
黃薑粉	3 湯匙
臘腸	150 克
臘肉	100 克

○糯米飯調味料

生抽	3.5 茶匙
蠔油	3.5 茶匙
鮮醬油	3.5 茶匙
老抽	半茶匙
麻油	半茶匙
熱開水	4.5 湯匙

8人份量

雙色米
脆豬

video 教學

Roast suckling pig stuffed with
bi-colour glutinous rice

做法

①雙色米飯

1. 紫番薯去皮，切幼粒，加入清水 600 克用攪拌機打爛，倒入盆內，加入糯米 450 克浸泡 1 晚（約 10 小時），即成紫色米，沖去表面澱粉，隔水備用。

2. 黃薑粉 3 湯匙與清水 600 克拌勻，倒入盆內，加入糯米 450 克浸泡 1 晚（約 10 小時），即成黃色米，隔水備用。

3. 雙色米分別隔水蒸 30 分鐘，取出，用溫水稍沖洗至鬆散，瀝乾水分。

4. 臘腸及臘肉隔水蒸 30 分鐘，切成幼粒，分成兩等份備用。

5. 用慢火燒熱鑊，先放入一份臘味料略炒熱至出油，加入紫色糯米飯及半份調味料炒勻，盛起備用。黃色米的做法相同。

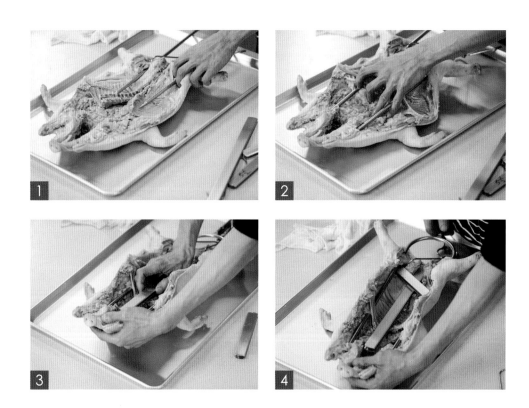

②釀乳豬及燒製

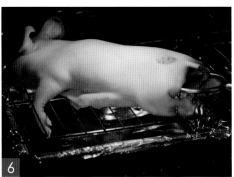

1. 乳豬平放，內籠向上，於尾龍骨位置用刀斬開缺口，沿着脊骨斬至頭部，取出豬腦。

2. 乳豬平放，內籠向上，慢慢去掉近頭部兩邊的首 4 條排骨，再起出膊骨，在厚肉位置略修薄，並在表皮剁上井形花紋。

3. 燒滾水，放入乳豬用慢火煮約 2 分鐘，取出，用清水浸至豬身冷卻，瀝乾水分。

4. 乳豬平放，內籠向上，將內籠搽勻醃料待 10 分鐘，串上豬叉及豬柴定型（圖 1 至 5）。

5. 豬皮用清水洗淨，抹乾，塗勻上皮料。

6. 焗爐預熱至 240℃，放入乳豬（皮向上），開門烤烘 25 分鐘，轉120℃再烘 45 分鐘，取出待冷。

7. 除了乳豬頭骨與大腿骨外，取出其餘的骨塊（圖 7 至 8）。

8. 將雙色糯米飯釀入豬內籠，用錫紙把飯包好，合上豬身，用鐵線將肚部縫合封口（圖 9 至 12）。

9. 乳豬放於豬叉中間，串上叉燒針，用鐵線綁實固定位置（圖 13 至 14）。

10. 燒熱油鑊至高溫，用熱油淋上乳豬皮炸至皮脆，再用火槍略燒至表面呈芝麻皮狀，去掉錫紙，切塊享用（圖 15 至 19）。

必學重點

- 斬開脊骨及起骨時,緊記小心處理,不可弄破豬皮。
- 乳豬塗抹上皮料後毋須風乾,可直接放入焗爐烘熟。
- 烘焗過程主要是將乳豬烘熟,毋須將表面烘至有色,如呈現金黃色的位置可用錫紙遮蓋,以免烘焦。
- 用熱油炸乳豬表皮時,如出現氣泡情況,需用針刺穿,讓空氣排出,以免過度膨脹。
- 由於乳豬內籠的骨全部去掉,故釀入糯米飯後,必須用豬叉及鐵線等固定位置,以免糯米飯散掉。

15

16

17

18

19

茄子香魚

Deep-fried grouper with
eggplant filling

6 人份量

材料

石斑魚	1 條（約 500 克）
茄子	半條
花雕酒	1 湯匙

醃料

醃味白滷水汁
300 克（做法參考 p13）

料頭

薑粒	1 湯匙
葱花	1 湯匙
蒜茸	1 湯匙

調味料

清水	200 克
大地魚粉	1.5 湯匙
回鍋醬	1 湯匙（做法參考 p.11）
豆瓣醬	1 茶匙
蠔油	1 茶匙
老抽	1 茶匙
鹽	1/8 茶匙

粟粉獻

粟粉	1 茶匙
清水	2 茶匙

做法

1. 石斑魚去鱗及內臟，由魚脊位置剝開魚肉，剪去中間大骨，形成空心狀，洗淨，放入醃料待 20 分鐘，抹乾水分備用。

2. 燒熱油鑊，將茄子插入魚身中空位置，用慢火炸 5 分鐘至魚身熟透呈空心的圓形，退出茄子，瀝乾油分，放於碟上。

3. 將茄子被炸乾的部位切除，取 120 克茄子切成約 1 吋 x 1 吋丁粒，放入滾油炸至金黃色，瀝乾。

4. 燒熱鑊，下油 1 湯匙爆香料頭，加入花雕酒 1 湯匙炒勻，下調味料及茄子丁煮滾， 轉慢火燜至茄子軟腍，埋獻，放入魚身內即可。

必 學 重 點

要將石斑魚炸至定型，必須將茄子按住緊貼鍋底，避免魚肉受熱收縮而脫離茄子。

○材 料

光雞	1 隻（約 2 斤 4 兩）
花雕酒	2 湯匙
芝麻	3 湯匙

○醃 料

幼鹽	2 湯匙
雞粉	1 湯匙
五香粉	1 湯匙

○上 皮 料

雞蛋白	1 個

○蘸 料

日式胡麻醬

脆 皮
芝 麻 雞

Crispy sesame chicken

做法

1. 光雞掏去肥油及內臟，洗淨，瀝乾水分。

2. 雞身用花雕酒塗勻後，再將已混合的醃料搽勻，醃 30 分鐘。

3. 雞平放蒸氣盆（胸肉向上），隔水蒸 40 分鐘，取出，用水略沖淨雞皮，掛起瀝乾水分。

4. 於雞皮塗勻上皮料，灑上芝麻待風乾。

5. 燒熱油至 7 成熱，炸至芝麻雞皮脆，斬件，伴蘸料享用。

必 學 重 點

雞皮塗勻上皮料後，必須即時灑上芝麻，此時黏合效果最佳，並須待十五分鐘，令蛋白與芝麻黏實，並進行風乾的步驟。

龍井
燻脆鴿

Deep-fried smoky squabs scented with
Longjing and chrysanthemums

○ 材 料

頂鴿　　　2 隻

○ 滷 水 料

清水　　　2 斤（1.2 公升）
五香粉　　1 湯匙
薑片　　　15 克
香葉　　　6 片
乾菊花　　15 克
龍井茶　　8 克（後下）

○ 滷 水 調 味 料

幼鹽　　　2 湯匙
黃糖　　　1 湯匙
雞粉　　　1 湯匙
花雕酒　　2 湯匙
煙燻液　　125 克

○ 上 皮 料

A 類上皮水　4 湯匙
（做法參考 p.13）

日本製的煙燻液

做法

1. 乳鴿掏去油膏及內臟,洗淨,剪去雙腳備用。

2. 乾菊花用煲魚湯袋盛好,放入滷水料內煮滾,轉慢火煮 5 分鐘,
   熄火,放入龍井茶葉浸泡 2 分鐘,隔渣,灑入幼鹽、黃糖及雞粉
   拌勻至完全溶解,待冷後加入花雕酒及煙燻液。

3. 乳鴿放入滾水內至滾,轉慢火煮 5 分鐘,熄火,加蓋焗 30 分鐘至
   熟透,放進滷水汁待 12 小時至入味(需存於雪櫃保鮮)。

4. 乳鴿瀝乾水分,塗勻上皮料及風乾,用熱油炸至皮脆即可。

必 學 重 點

要待滷水汁完全冷卻後,才加入花雕酒及煙燻液,因熱滷水
汁會使酒香及煙燻味揮發掉。

鳳城四寶紮

Grilled cabbage rolls with four-treasure filling

6 人份量

◯材料

鮮蝦肉	6 隻
豬頸肉	半件
煙鴨胸	半件
紫番薯	50 克
幼露筍	6 條
洋葱	半個
芫茜	1 棵
椰菜葉	6 片
娃娃菜	12 片
豬腩片	6 片（每片約 2 厘米厚）
蜜糖	10 湯匙

◯醃料

砂糖	5 湯匙
幼鹽	1 湯匙
雞粉	半湯匙
海鮮醬	1 湯匙
芝麻醬	1 湯匙
蠔油	2 湯匙
花雕酒	1 湯匙
雞蛋	1 個
蒜茸	1 湯匙
乾葱茸	1 湯匙
薑茸	2 湯匙

預備做法

1. 豬頸肉洗淨，加入醃料待 30 分鐘，放入焗爐用 240℃兩面各烘 8 分鐘，轉 120℃烘 15 分鐘。

2. 煙鴨胸解凍，用焗爐 180℃烘 10 分鐘。

3. 椰菜葉用熱水慢火煮約 5 分鐘至軟身。

必 學 重 點

- 如買不到豬腩片，可將整件五花腩肉冷藏至硬身，去皮後切成薄片。
- 豬腩片包捲菜包時，不要留有空隙，否則烘焗時腩片易乾硬。

做法

1. 鮮蝦肉與醃料拌勻，醃 10 分鐘。

2. 將熟豬頸肉、煙鴨胸、紫番薯、露筍、洋葱及芫茜分別切成約 3 吋長條形（如筷子般），分配成 6 等份。

3. 椰菜葉鋪平，疊上一片娃娃菜，放上蝦肉及一份餡料，蓋上另一片娃娃菜，把椰菜葉包好。

4. 豬腩片塗勻醃料，沿着椰菜葉緊緊包捲，用牙籤封口固定，置於焗盆內。

5. 焗爐預熱至 240℃，放入四寶紮每面各烘 10 分鐘，轉 150℃再烘 10 分鐘，取出塗勻蜜糖，再用 240℃烘至兩面蜜糖沸騰即成。

牛骨茶

Beef Bak Kut Teh
(Beef bone herbal soup)

6 人份量

○ 材料

牛肋骨	1 斤

○ 滷水料

清水	700 克	玉竹	6 克
當歸	3 克	黨參	5 克
參根	5 克	圓肉	2 粒
北芪	3 克	蒜頭	3 大個
肉桂	1.5 克	生抽	3 湯匙
八角	2 粒	蠔油	1.5 湯匙

○ 蘸料

燒味汁	2 湯匙（做法參考 p.10）

做法

1. 牛肋骨切去表面多餘的肥膏，放入滾水略燙 2 分鐘，用清水沖淨，備用。

2. 煮滾滷水料，加入牛肋骨煲滾，轉慢火加蓋煮約 90 分鐘，直至肉質軟腍即成。

必 學 重 點

- 煮牛肋骨時不宜用大火，否則肉質容易鬆散，慢火煮才嫩滑。
- 如將豬肋骨替代牛肋骨，即是傳統的肉骨茶做法。

欖角
燒花腩

Grilled pork belly in
preserved black olive sauce

◯ 材料

五層腩	2 斤	蜜糖	6 湯匙
生粉	6 湯匙	燒味汁	4 湯匙
			（做法參考 p.10）

◯ 醃料

砂糖	5 湯匙	老抽	半茶匙
幼鹽	1 湯匙	欖角茸	3 湯匙
雞粉	1 茶匙	豆豉茸	1 湯匙
海鮮醬	2 湯匙	玫瑰露酒	1 茶匙
蠔油	2 湯匙	蒜茸	1 湯匙
芝麻醬	半湯匙	乾葱茸	半湯匙
生抽	半茶匙		

做法

1. 五層腩去皮，修平底部腩肉，切成約 1.5 吋
   闊長方條，用 3 湯匙生粉調勻少許水，塗勻
   腩肉條醃 4 小時（需儲存雪櫃保鮮）。

2. 腩肉條洗淨，再用生粉漿塗勻，放入醃料內
   待 80 分鐘，盛起。

3. 焗爐預熱至 240℃，放入腩肉條每面烘至出
   現微黑焦糖後，轉 100℃烘 30 分鐘，取出。

4. 腩肉條塗勻蜜糖，再用 240℃烘至每面蜜糖
   沸騰後，轉用 100℃烘約 20 分鐘，取出，最
   後塗抹蜜糖。

5. 享用時，將腩肉切片，淋上燒味汁即可。

必學重點

不要選帶羶味的五花腩，影響烘製出來的香味。

video 教學

○ 材料

豬肚	1 個
五花腩	600 克
臘肉	100 克
花生	50 克
蓮子	50 克

○ 滷水材料

清水	2 公斤
生抽	100 克
黃糖	100 克
老干媽油辣椒	2 湯匙
豆瓣醬	1 湯匙
幼鹽	1/4 茶匙
老抽	4 湯匙
南乳	45 克
海鮮醬	3 湯匙
白豆蔻	12 粒
桂皮	3 克
八角	2 粒
沙薑粒	6 粒
白胡椒粒	1 茶匙
五香粉	2 茶匙

○ 滷水料頭

油	2 湯匙
薑片	40 克
乾葱頭	60 克（切片）
蒜肉	60 克（切片）
花雕酒	1 湯匙（後下）

○ 埋獻料 A

生粉	半茶匙
清水	1 茶匙

○ 埋獻料 B

生粉	1 茶匙
清水	2 茶匙

金銀羅漢肚

Marinated stuffed pork tripe

做法

1. 花生及蓮子預先用清水浸泡 2 小時，盛起備用。

2. 豬肚剪去表面油脂，反轉內壁向外，將粗鹽、生粉及白醋分成三份，依次揉擦豬肚，每次揉擦後用清水沖淨，重覆三次，直至去除肚內的黏液及異味，剪掉黃色內膜，沖水後，將內壁反轉向內。

3. 五花腩切成約手指般大的長方塊，放入滾水用慢火煮 10 分鐘，盛起備用。

4. 滷水料頭爆香後，加入滷水材料煮滾，下五花腩、花生及蓮子，轉慢火加蓋煲 90 分鐘。

5. 臘肉放入蒸籠，隔水蒸 30 分鐘，切成薄片。

6. 用麻繩將豬肚的細肚口位綁緊，打開大肚口位，釀入五花腩、臘肉、花生及蓮子，取滷水汁 100 克與預先混合的埋獻料 A 煮滾，倒入豬肚內，綁緊豬肚口。

7. 豬肚放入滷水內煮滾，加蓋，轉慢火煲 1 小時（須每隔 15 分鐘轉一次），使豬肚兩面均勻入味。

8. 取滷水 160 克與預先混合的埋獻料 B 煮滾，淋在已剪開的豬肚上，即可享用。

必 學 重 點

豬肚必需徹底洗淨，否則煮熟後帶有異味，影響滷水的香味。

Drunken squab in Shaoxing wine with dried plums

Makes 6 servings

Tips:
Make sure the squabs are soaked in the marinade long enough for a robust flavour. If time allows, try to steep them in the marinade for 24 hours for the best results.

Ingredients:
450 g squabs

Shaoxing wine with dried plums:
120 g premium Shaoxing wine
2 liquorice-scented dried plums
1/2 tbsp candied osmanthus

Spiced marinade:
900 g water
2 whole pods star anise
4 bay leaves
3 sand ginger roots
1 tbsp cumin
15 g sliced ginger
4 liquorice-scented dried plums

Seasoning for marinade:
70 g table salt
30 g chicken bouillon powder
80 g premium Shaoxing wine

Method:
1. Make Shaoxing wine with dried plums by mixing all ingredients together. Let the dried plums steep in the wine for 2 days for flavours to infuse.
2. Boil the spiced marinade. Turn to low heat and simmer for 10 minutes. Turn off the heat. Let cool and add seasoning. Stir until salt dissolves.
3. Remove innards and fat inside the squabs. Rinse well.
4. Boil a pot of water and put in the squabs. Bring to the boil again and turn to low heat. Cook for 5 minutes. Turn off the heat. Cover the lid and leave them in the water for 30 minutes until cooked through. Transfer the squabs into the spiced marinade. Let them steep in the marinade for 3 hours.
5. Before serving, chop the squabs into pieces. Drizzle with some Shaoxing wine with dried plums from step 1. Serve.

Bang Bang chicken

(Boneless chicken and mung bean sheet noodles dressed in Sichuan spicy sesame sauce)

Makes 6 servings

Ingredients:
- 1/2 dressed chicken (about 700 g)
- 600 g fresh mung bean sheet noodles
- 150 g cucumber
- 30 g Peking scallions

Dressing:
- 2 tsp sugar
- 1/2 tsp table salt
- 1/2 tsp chicken bouillon powder
- 1 tsp Sichuan pepper oil
- 4 tsp Zhenjiang black vinegar
- 2 tbsp sesame oil
- 2 tbsp light soy sauce
- 2 tbsp sesame paste
- 2 tbsp Lao Gan Ma brand chilli oil
- 2 tbsp grated garlic
- 6 tbsp coriander (cut into short lengths)

Method:
1. Mix the dressing well.
2. Remove innards of the dressed chicken. Rinse well.
3. Boil a pot of water and put in the chicken. Bring to the boil again. Turn to low heat and simmer for 5 minutes. Turn off the heat and cover the lid. Leave the chicken in the water for 30 minutes until cooked through. Remove chicken from the water. Dunk in cold drinking water to let cool.
4. Cut the mung bean sheet noodles into 1/2-inch-wide ribbons. Blanch in boiling water until they turn transparent. Drain and soak in cold drinking water until cooled. Drain and set aside.
5. Debone the chicken. Cut chicken, cucumber and Peking scallions into thick strips. Arrange on a serving plate. Stir in the mung bean noodles and dressing. Serve.

Tips:
Make sure you control the time well when cooking the mung bean noodles. Drain them when the start to turn transparent. Otherwise, they may shrink and turn mushy if overcooked.

Pickled cucumber in aged vinegar and garlic dressing

Makes 6 servings

Ingredients:
900 g hothouse baby cucumbers

Pickling brine:
70 g Shanxi aged vinegar
10 g light soy sauce
30 g sesame oil
30 g spicy bean sauce
50 g garlic cloves (crushed)
10 g coriander (cut into short lengths)
2 tbsp deep-fried grated garlic
50 g sugar
10 g table salt
10 g chicken bouillon powder
1 tsp Sichuan pepper oil

Method:
1. Rinse the cucumbers. Drain and cut off both ends. Crush them with the flat side of a knife so that they crack into strips. Remove the seeds and cut into 2.5-inch-long pieces.
2. Mix the pickling brine well. Put in the cucumber and toss well. Leave it for 1 hour. Serve.

Tips:
- *It's advisable to make the pickling brine 1 hour ahead. That would allow enough time for the garlic flavour to infuse.*
- *After putting the cucumber into the pickling brine, stir it every 15 minutes. That would make sure every strip of cucumber picks up the pickling brine evenly.*

Pickled young ginger

Makes 6 servings

Ingredients:
1.2 kg young ginger
80 g table salt

Pickling brine:
900 g water
900 g rock sugar
30 g table salt
600 g Chaoshan rice vinegar
3 slices lemon

Method:
1. To make the pickling brine, boil 900 g of water. Turn to low heat and add rock sugar and table salt. Stir until they dissolve. Turn off the heat and let cool. Pour in rice vinegar and put in the lemon. Stir and set aside.
2. Rinse the ginger. Peel it and cut into slices about the thickness of a chopstick. Sprinkle with 80 g of table salt. Stir well and let it sit for 45 minutes. Rinse in cold running water for 90 minutes to remove all saltiness.
3. Wipe dry the ginger with a clean towel. Put ginger into the pickling brine from step 1. Put a heavy ceramic bowl or dish over the ginger to make sure every piece is soaked in the brine. Leave them for at least 2 days.

Tips:
- *If you use white vinegar in place of rice vinegar, make sure you taste the pickling brine after making it. It shouldn't be too sour. Otherwise, the brine will pull too much water out of the ginger drying it out and making it chewy. If the brine is too sour, thin it out with water and add some more rock sugar.*
- *Store the pickled ginger in a fridge to keep it fresh. It's also crispier when served chilled.*

Chilled marinated abalones

Makes 6 servings

Tips:
- *Soaking live abalones in warm salted water helps remove the dirt on them.*
- *When you cook the abalones, do not start with high temperature. Otherwise, the abalones will shrink too quickly and may crack. That would adversely affect the presentation and the mouthfeel.*

Ingredients:
- 6 live baby abalones
- 20 g sliced ginger
- 20 g spring onion

Spiced marinade:
- 300 ml water
- 10 g galangal
- 1 stem lemongrass
- 10 g ginger
- 2 bay leaves
- 2 slices sand ginger
- 1 slice dried liquorice
- dried tangerine peel
- 1/2 red chilli

Seasoning for marinade:
- 18 g rock sugar
- 9 g table salt
- 2 g chicken bouillon powder

Method:
1. Add 1 tsp of salt to 300 ml of warm water (50°C). Stir well. Put in the abalones to soak for 10 minutes. Drain and rinse in water for a few minutes.
2. Boil the spiced marinade. Cover the lid and cook over low heat for 8 minutes. Add seasoning and stir until dissolves. Turn off the heat. Let cool.
3. In a pot, heat water to 50°C. Add sliced ginger, spring onion and abalones. Cook over low heat until it boils. Keep on boiling for 2 minutes till abalones are cooked through. Shell the abalones. Remove the innards. Rinse well.
4. Wipe dry the abalones. Soak them in the cooled spiced marinade for 5 hours. Serve chilled.

Hand-shredded chicken dressed in wasabi sauce

Makes 6 servings

Ingredients:
1 piece chicken breast (about 200 g)
40 g celery
40 g carrot
30 g red bell pepper
30 g yellow bell pepper
20 g Peking scallions
20 g pickled young ginger
1 tsp toasted sesames
1 tbsp flying fish roe

Dressing:
4 tsp oyster sauce
4.5 tsp sugar
1 tsp sesame oil
1/4 tsp chicken bouillon powder
1/2 tsp dark soy sauce
1 tsp wasabi (Japanese horseradish)

Method:
1. Boil a pot of water. Put in the chicken breast and bring to the boil again. Turn off the heat and cover the lid. Leave the chicken breast in the hot water for 15 minutes until cooked through. Remove the chicken breast and soak it in cold drinking water until completely cooled. Tear into shreds with your hands.
2. Rinse and finely shred the celery, carrot, bell peppers and Peking scallions into matchsticks. Shred the young ginger into matchsticks too. Add all shredded ingredients to the shredded chicken from step 1. Mix all dressing ingredients. Pour dressing over the chicken mixture and toss well. Sprinkle with toasted sesames and flying fish roe at last. Serve.

Tips:
- *Do not boil the chicken over high heat. It should be poached in barely boiling water to keep the flesh juicy and tender.*
- *After tossing all ingredients in the dressing, serve immediately. Otherwise, moisture will be drawn out of the ingredients by the dressing, thinning it out and vegetables will wilt and turn soggy.*

Thousand-layer pork ear terrine

Makes 6 servings

Tips:
- *When you roll the pork ears in a bamboo mat, try to evenly lay the pork ears on the top and middle portion of the mat. Leave the bottom 1 inch of the mat uncovered. When you roll it up, do it as tightly as you can. Otherwise, there might be empty spaces in the terrine after set and the terrine may come apart easily.*
- *It's advisable to trim off the thickest part of the pork ears. They tend to overlap and adhere to each other better when rolled and pressed.*

Ingredients:
  6 large pork ears
  1 sprig spring onion
  5 slices ginger
  1 tsp ground white pepper
  40 g gelatine powder
  200 ml water

Spiced marinade A:
  1 tbsp oil
  20 g sliced garlic
  30 g sliced shallot
  30 g spring onion (cut into short lengths)
  1/2 tbsp Sichuan peppercorns
  2 whole pods star anise
  1 red chilli
  1 tbsp premium Shaoxing wine (added last)

Spiced marinade B:
  1.3 litre water
  100 g light soy sauce
  20 g dark soy sauce
  7 g table salt
  75 g raw cane sugar slab
  10 g chicken bouillon powder
  70 g oyster sauce
  1/4 tsp ground white pepper
  20 g galangal
  15 g coriander

Method:
1. Burn off the hair on the pork ears. Rinse well. Trim off the fat.
2. Boil 1.5 litre of water in a pot. Put in spring onion, sliced ginger, ground white pepper and pork ears. Cook over low heat for 15 minutes to remove the gamey taste. Rinse in cold water.
3. Put a bamboo mat on the bottom of the pot. Put in the pork ears. Stir fry spiced marinade A ingredients until fragrant. Add spiced marinade B ingredients. Bring to the boil. Pour it into the pot and bring to the boil again. Turn to low heat and cook for 2 hours. Turn off the heat.
4. In a bowl, add gelatine powder to 200 ml of water. Wait till the gelatine is soaked through. Put the bowl on a pot of simmering water. Stir until gelatine dissolves. Add the gelatine solution to the spiced marinade. Mix well.
5. Lay flat a bamboo mat for sushi roll on the counter. Brush sesame oil on it. Stack the pork ears over each other and form the rectangular shape. Roll up a bamboo mat and tie it tight with hemp strings. Put some heavy weight over it. Refrigerate for 8 hours. Cut into slices and serve chilled.

Deep-fried sweet potato balls with pork floss filling

Makes 12 croquettes

Ingredients:
- 170 g sweet potato
- 125 g glutinous rice flour
- 30 g flour
- 100 ml hot water
- 1.5 tbsp brown sugar
- 3 tsp black sesames
- 7 tsp white sesames

Filling:
- 4 tsp peanut butter
- 12 tsp pork floss

Method:

1. Rinse, peel and cut sweet potato into pieces. Steam for 20 minutes until done. Let cool.
2. Add brown sugar to hot water and stir until it dissolves. Put in the sweet potato and mash it. Let cool. Add glutinous rice flour and flour. Knead into dough.
3. Divide the dough into 12 equal pieces. Roll them round and press into a round patty. Put 1/4 tsp of peanut butter and 1 tsp of pork floss on each patty. Fold the edges toward the centre to wrap the filling. Seal the seam and roll them round. Roll them in black and white sesames to coat evenly.
4. Pour cold oil in a wok. Put in the sweet potato balls and heat over low heat until hot. Deep fry the sweet potato balls until they float and puff up. Turn to high heat and fry until golden brown. Drain and put on paper towel to remove excessive oil. Serve.

Tips:

- *Sweet potatoes may contain different amount of water, which may affect the dampness of the dough. If it's too dry or stiff, add a little water at a time to soften it.*
- *After you wrap in some filling, if the sweet potato balls start to crack, you may dip your finger in some water and gently rub on the crack to patch it. It also makes the sweet potato balls damper and easier for the sesames to adhere.*
- *When you deep-fry the sweet potato balls, it's essential to control the oil temperature. Start with low heat. Otherwise, the surface of the sweet potato balls may be stiffened by high heat and fail to puff up later on. That would result in sweet potato balls that are too hard and dense. Wait patiently until they puff up and float before turning up the heat to deep-fry until crispy.*

Honey-glazed pork jerky

Makes 6 servings

Ingredients:
300 g half-fatty ground pork
1 sheet aluminium foil

Seasoning:
4 tsp multifloral honey
1/2 tsp table salt
1/2 tsp chicken bouillon powder
1/2 tsp light soy sauce
1 tsp oyster sauce
1 tsp Hoi Sin sauce
1/2 tsp Chinese rose wine
1/2 tsp five-spice powder

Method:
1. Put ground pork into a large bowl. Add seasoning and mix until sticky. Refrigerate for 2 hours.
2. Line a baking tray with aluminium foil. Evenly spread the ground pork on it. Cover in cling film. Roll the pork out until 2 mm thick uniformly with a rolling pin. Remove cling film.
3. Preheat an oven to 175°C. Bake the pork for 8 minutes on each side. Drain the liquid on the baking tray if any. Then put the pork back in and bake on each side for 5 more minutes.
4. Bake till the pork jerky glistens with oil and lightly browned. Remove from oven and let cool. Serve.

Tips:
- *Make sure you stir the ground pork until sticky and roll it flat. That way, the pork will bind better into one sheet after baked.*
- *The pork should be ground finely. If the pork is chunky, it's hard to dry it properly and it won't taste like pork jerky at last.*

Grilled chicken wings stuffed with durian and cheese

Makes 6 servings

Ingredients:
- 6 whole chicken wings
- 120 g durian flesh
- 30 g cheddar cheese
- 3 tbsp honey

Marinade:
- 5 tbsp sugar
- 1 tbsp table salt
- 1/2 tbsp chicken bouillon powder
- 2 tbsp Hoi Sin sauce
- 1 tbsp oyster sauce
- 1 tbsp sesame paste
- 1 tbsp dark soy sauce
- 1 tbsp light soy sauce
- 1 tbsp grated ginger
- 1 tbsp finely chopped shallot
- 1 tbsp grated garlic

Method:
1. Rinse the chicken wings and debone them. Stuff 20 g of durian flesh and 5 g of cheese into each chicken wing. Seal the seam securely with a toothpick.
2. Mix the marinade well and pour in with the stuffed chicken wings. Leave them for 30 minutes.
3. Preheat an oven to 200°C. Grill the chicken wings for 5 minutes on each side. Brush honey on them. Then grill for 5 minutes more on each side. Serve.

Tips:
- *Try not to pierce through the skin when you debone the wings. Otherwise, the filling may ooze out and drip into your oven.*
- *Do not over-stuff the wings with durian. The durian flesh will expand in the grilling process and the wings may burst if over-stuffed.*

Paprika scented grilled pork cheek

Makes 6 servings

Tips:

Do not bake the pork for too long over high temperature. Otherwise, that may dry it out and the pork may burn.

Ingredients:

2 pork cheeks
2 tbsp honey
1.5 tbsp Chinese rose wine

Marinade:

5 tbsp sugar
1 tbsp table salt
1 tsp chicken bouillon powder
1.5 tsp five-spice powder
2 tsp paprika
1 tsp ground coriander seeds

Dip:

4 g hot water
30 g golden syrup
16 g dark soy sauce
8 g light soy sauce
4 g lime juice
8 g white part of spring onion (cut into short lengths)
8 g coriander (cut into short lengths)
1 bird's eye chilli (finely chopped)

Method:

1. Rinse the pork and drain. Add Chinese rose wine and mix well. Mix the marinade well and add to the pork. Rub evenly and let it sit for 30 minutes.
2. Preheat an oven to 240°C. Put in the pork and grill each side for 10 minutes. Brush on honey. Turn the oven down to 200°C and bake each side for 5 more minutes. Remove from oven to let cool.
3. Before serving, slice pork cheek thinly at an angle. Drizzle with the mixed dip. Serve.

Crispy honeycomb tofu

Makes 6 servings

Ingredients:
  2 cubes cloth-wrapped tofu

Spiced marinade:
  600 g water
  2 whole pods star anise
  10 sand ginger roots
  1 tsp clove
  5 bay leaves
  5 g cassia bark
  1/2 tsp ground white pepper
  2 sprigs coriander (cut into short lengths)
  40 g Chinese celery (cut into short lengths)
  1 red chilli (cut into short lengths)
  10 slices ginger

Seasoning for marinade:
  2 tbsp table salt
  2 tsp sugar
  1 tsp chicken bouillon powder

Deep-frying batter:
  60 g flour
  2 g wheat starch
  2 g potato starch
  7 g baking powder
  45 g water
  45 g oil

Method:
1. Boil the spiced marinade in a pot. Turn to low heat and cover the lid. Cook for 10 minutes. Turn off the heat. Add seasoning for marinade. Stir until sugar dissolves. Set aside.
2. Cut each cube of tofu into quarters. Steam for 5 minutes. Soak tofu in the marinade for 1 hour. Drain well.
3. Heat oil in a wok. Deep-fry the tofu until golden. Drain and let cool.
4. Mix the deep-frying batter. Dip each piece of tofu in the batter to coat well on all sides.
5. Heat oil in a wok again. Deep-fry tofu until golden and crispy. Serve.

Tips:
- *Make sure the oil is smoking hot before putting in the tofu. Otherwise, there will be blisters on the crust and the tofu won't appear smooth.*
- *After putting tofu into the hot oil, make sure the crust is stiffened before flipping it. Otherwise, it's hard to keep it in one piece.*

Deep-fried pork intestines stuffed with minced cuttlefish

Makes 6 servings

Ingredients:
600 g pork large intestines
200 g cuttlefish (dressed)

For cleaning intestines:
120 g coarse salt
120 g potato starch

Soup for blanching intestines:
200 g of water
3.5 tbsp white vinegar
40 g ginger
40 g spring onion (cut into short lengths)
3 whole pods star anise
1 tsp ground white pepper

Marinade for minced cuttlefish:
1/4 tsp salt
1/4 tsp chicken bouillon powder
2 tsp cornstarch
1 tsp chopped coriander stems

Spiced marinade:
1.5 kg water
15 g oyster sauce
30 g salt
20 g rock sugar
12 g chicken bouillon powder
3 whole pods star anise
3 g cassia bark
1 tbsp Sichuan peppercorns
3 Tsaoko fruits
1 tsp ground white pepper
60 g chopped shallot
20 g sliced ginger
1 tsp five-spice powder
1 tbsp premium Shaoxing wine
1 tbsp Chinese rose wine
1/2 tbsp red yeast rice

Basting sauce:
35 g water
35 g white vinegar
10 g maltose
5 g red vinegar
20 g double-distilled Chinese rice wine

Sweet and sour dip: (mixed well)
1 tbsp red vinegar
1 tbsp brown sugar
1 tbsp plum sauce
1 tbsp ketchup

Method:

Cleaning the intestines:

1. Flip the intestines inside out. Rub coarse salt and flour all over to remove slime and dirt. Rinse with water. Repeat the rubbing and rinsing steps three times. Then flip the intestines back in. Rinse with water again.
2. Boil the soup for blanching intestines. Put in the intestines and heat until it boils. Turn to low heat and simmer for 80 minutes. Drain. Rinse the intestines with water again. Drain well.

Minced cuttlefish:

1. Rinse the cuttlefish. Wipe dry. Cut into small pieces. Put into a food processor and grind it.
2. Put the ground cuttlefish into a bowl. Add seasoning and stir in one direction until sticky.

Marinating the intestines:

1. Boil the spiced marinade. Put in the pork intestines. Cook over low heat for 30 minutes. Take the intestines out and rinse with hot water. Hang them up and let dry.
2. Heat the basting sauce in a pot until maltose melts.
3. Brush the basting sauce all over the intestines. Let them air dry.
4. Stuff the intestines with minced cuttlefish. Heat oil in a wok up to 200°C. Put in the stuffed intestines and deep fry until crispy. Slice and serve with the sweet and sour dip on the side.

Five-spice pork roll in Minnan style

Makes 6 servings

Tips:

Make sure you steam the beancurd sheet roll for the right period of time. Do not over-steam it. Otherwise, the beancurd sheet may break down adversely affecting the presentation and the texture of this dish.

Ingredients:
1 beancurd sheet
200 g pork belly
60 g water chestnut (peeled)
40 g onion
40 g shallot
40 g white part of spring onion
5 tbsp sweet potato flour
2.5 tbsp flour
1 egg

Marinade:
1 tbsp light soy sauce
1/2 tsp sugar
2 tsp premium Shaoxing wine
1/2 tsp table salt
1/2 tsp five-spice powder

Flour batter (mixed well):
1 tbsp flour
1.5 tbsp water

Method:
1. Rinse the pork. Finely dice pork, water chestnut, onion, shallot and white part of spring onion so that they resemble the size of a soybean. Add marinade and stir well. Leave it for 15 minutes. Add sweet potato starch, flour and egg. Stir and knead until sticky. This is the pork filling.
2. Cut the beancurd sheet into 4 equal pieces. Brush hot water on them to soften.
3. Divide the pork filling from step 1 into 4 equal portions. Put one portion on each beancurd sheet. Roll it up firmly. Seal the seam with flour batter.
4. Brush oil on a steaming tray. Put the beancurd sheet rolls on it and steam for 25 minutes. Remove from heat and let cool.
5. Slice the beancurd sheet rolls into 1/2-inch-thick pieces. Deep-fry in oil until golden and crispy. Serve hot.

Ingredients:
12 chicken feet

Abalone sauce:
1.4 litre water
60 g oyster sauce
20 g sugar
120 g dark soy sauce
1 tbsp premium Shaoxing wine
1 tsp five-spice powder
4 bay leaves
3 whole pods star anise
30 g dried shrimps
30 g dried plaice
20 g sliced ginger
20 g spring onion (cut into short lengths)
20 g shallot (sliced)

Method:
Abalone sauce:
1. Heat oil in a wok. Deep-fry sliced ginger, shallot, spring onion, dried shrimps and dried plaice until golden. Transfer to the pot.
2. Add the remaining abalone sauce ingredients to the fried mixture from step 1. Bring to the boil over low heat. Turn off the heat and set aside.

Braised chicken feet:
1. Rinse the chicken feet and trim off the nails. Blanch in boiling water for 2 minutes. Drain and soak in cold water until completely cooled. Drain and set aside.
2. Bring the abalone sauce to the boil. Put a bamboo mat on the bottom of the pot. Put in the chicken feet and bring to the boil again. Turn to low heat and simmer for 45 minutes. Serve.

Chicken feet braised in abalone sauce

Makes 6 servings

Tips:
The abalone sauce is originally used to stew abalones and it can be re-used. Just re-boil it after use and let cool. Then keep the leftover in the fridge for other uses, such as dressing E-fu noodles or braising chicken feet.

Taro balls in baby oyster soup

Makes 6 servings

Ingredients of taro balls:
- 200 g taro
- 80 g sweet potato flour

Seasoning for taro balls:
- 2 tbsp hot water
- 1/4 tsp salt
- 1/4 tsp chicken bouillon powder

Aromatics for soup:
- 1.5 tbsp oil
- 25 g sliced ginger
- 25 g sliced shallot
- 20 g sliced galangal

Seasoning for soup:
- 1 tsp salt
- 1/2 tsp chicken bouillon powder

Soup base A:
- 600 g water
- 100 g taro (finely grated)
- 100 g baby Napa cabbage
- 18 taro balls

Soup base B:
- 200 g baby oysters
- 1 sprig coriander (cut into short lengths)
- 1 sprig spring onion (cut into short lengths)
- 1 leek (cut into short lengths)

Method:
Taro balls:
1. Grate the taro finely. Add sweet potato starch. Mix seasoning for taro balls well and add to the taro mixture. Knead into dough. Squeeze between your thumb and index finger into small balls about 15 g each. You need 18 taro balls for 6 servings.
2. Brush oil on a bamboo steamer or steaming tray. Put the taro balls on it and steam for 15 minutes.

Baby oysters and mash taro soup:
1. Heat a wok and stir-fry aromatics for soup until fragrant. Add soup base A ingredients and bring to the boil. Stir in seasoning. Turn to low heat and cook for 5 minutes.
2. Add pre-wash soup base B ingredients and bring to the boil again. Cook for 2 more minutes. Serve.

Tips:
- *Do not overcook the baby oysters. Just cook until they are just done. Otherwise, they may turn rubbery and bland.*
- *After squeezing the taro balls, steam them first before adding to the soup. If you add raw taro balls directly to the soup, the sweet potato flour in the taro balls would stick to the bottom of the pot and burn.*
- *Before squeezing taro balls with your hands, grease them with oil first. Otherwise, the mashed taro would stick to your hands and it'd get really messy.*

Ingredients:
1 dressed chicken (about 1.2 kg)
1 dried lotus leaf (large)
2 sheets aluminium foil (30 X 35 cm each)

Marinade A:
1 tbsp dark soy sauce
3 tsp table salt
2 tsp sugar
1 tsp chicken bouillon powder
1 tsp five-spice powder

Marinade B:
4 tbsp grated fresh sand ginger
4 tbsp finely chopped shallot
4 tbsp grated ginger
1/2 tbsp light soy sauce
4 tbsp premium Shaoxing wine

Filling:
1 tbsp oil
50 g fatty pork (cut into cubes)
30 g onion (diced)
1 whole pod star anise
15 g garlic cloves
3 dried shiitake mushrooms (cut into strips)
50 g chicken leg mushrooms (cut into strips)
6 small round chestnuts
1/2 tsp sesame oil

Seasoning for filling:
100 g water
3 tsp dark soy sauce
1 tsp light soy sauce
1 tsp oyster sauce
table salt

Salt crust:
120 g egg white
650 g table salt

Thickening glaze:
1 tbsp potato starch
2 tbsp water

Beggar's chicken
(roast chicken in salt crust)

Makes 8 servings

Method:

1. Soak the lotus leaf in hot water for 30 minutes. Drain and set aside.
2. Remove innards and fat from inside the chicken. Rinse well. Break the neck with the back of a knife. Tap the back of the chicken to flatten a little. Then flip the breast side up and press firmly to flatten. Cut the skin between the thigh and the breast and spread the legs and flatten the chicken further.
3. Brush the dark soy sauce from marinade A over the chicken skin. Then add the remaining marinade A and rub evenly on both the insides and outsides. Leave it for 40 minutes. Rinse off the marinade and drain. Shallow-fry in a wok until the skin is golden.
4. Put the chicken into a tray with the breast side up. Mix the marinade B well. Rub it on both the insides and outside of the chicken. Leave it for 30 minutes.
5. Stir fry the filling until fragrant. Add seasoning for filling and cook over low heat for 10 minutes with the lid covered. Stir in thickening glaze and cook till it thickens. Stuff the chicken with the hot filling. Wrap it in lotus leaf and then in two sheets of aluminium foil.
6. Mix the salt crust well. Spread a little on the bottom of the baking tray. Bake in a preheated oven at 200°C for 3 minutes till set. Put the chicken wrapped in aluminium foil over it. Then smear the remaining salt crust mixture evenly all over to seal well.
7. Preheat an oven to 200°C. Bake the chicken for 70 minutes. Crack open the salt crust. Remove the aluminium foil and lotus leaf. Chop up the chicken and serve.

Tips:

- *After you marinate the chicken for 40 minutes. Rinse off the marinade. Otherwise, the chicken may taste too salty.*
- *To make the salt crust, it's advisable to beat the egg white until stiff before mixing in the salt.*
- *Traditionally, "Beggar's chicken" is roasted in a crust of clay. This modern version cooks more quickly so that the chicken is kept tender and succulent.*

Ingredients:

300 g pork fat, 300 g pork rib eye
10 large chicken livers, 15 slices ginger
12 slices pickled young ginger (on the side)
12 lettuce leaves (on the side)

Marinade:

5 tbsp sugar, 1 tbsp table salt
1/2 tbsp chicken bouillon powder
1/8 tsp five-spice powder, 2 shallots (finely chopped)
3 cloves garlic (grated), 1/2 tbsp light soy sauce
1/2 tbsp dark soy sauce
1/2 cube red fermented tarocurd
1/2 tbsp Chu Hau sauce, 1/2 tbsp sesame paste
1 tbsp Chinese rose wine, 1 egg

Preparation:
Pork fat:

1. Cut the pork fat into a cube measuring 4 cm on each side. Blanch in boiling water for 10 minutes. Rinse in water until cooled. Cut thinly into slices about 0.3 cm thick. Drain. Add 5 tbsp of sugar and mix well. Leave it for at least 10 hours.
2. Add 1 tbsp of Chinese rose wine 20 minutes before baking in an oven.

Pork rib eye:

Leave the pork rib eye into the freezer until slightly stiffened. Then cut into slices about the same size as the pork fat. Add 2 tbsp of potato starch and 2 tbsp of water. Mix well and leave it for 30 minutes.

Chicken livers:

Add 2 tbsp of premium Shaoxing wine and 2 tbsp of grated ginger. Mix well and leave them for 30 minutes.

Assembly:

1. Mix all marinade ingredients. Then put in the pork fat, pork rib eye and chicken liver. Mix well and leave them for 20 minutes. Put them on metal or bamboo skewers alternately in the following order: sliced ginger, pork rib eye, chicken liver, pork fat, then sliced ginger again.
2. Preheat an oven over the highest heat up to 250°C. Grill the skewed meat for 5 minutes until lightly caramelized. Turn the oven down to 125°C and keep grilling for 20 minutes. Brush honey on them.
3. Turn the heat up again to 250°C. Grill until the honey bubbles (for about 3 minutes). Turn the oven down to 100°C and bake for 15 minutes. Take the skewers out and brush on honey.
4. Trim the lettuce leaves into round cups. Put a slice of pickled ginger in each lettuce cup. Remove the chicken liver and pork from the skewers. Arrange in the lettuce cups. Serve.

Grilled chicken liver and pork medallions

Makes 6 servings

Tips:

- *The pork fat used in this recipe is from the first subcutaneous fat layer behind a pig's ear. It is firmer in texture and holds it shape well after prolonged cooking. It is also crunchier in texture.*
- *Marinating the pork fat for 10 hours in sugar makes the pork fat even crunchier after grilled.*
- *When you put the meats on the metal or bamboo skewers, do not overcrowd them. Otherwise, the meats will be hard to cook through.*

Ginger duck soup

Makes 8 servings

Ingredients:
1/2 Muscovy duck (about 1.5 kg)

Aromatics for ginger soup:
3 tbsp black sesame oil
45 g sliced ginger
2 whole pods star anise
80 g rice wine (added last)

Ginger soup base:
500 g water
40 g sliced ginger
2 g Dang Gui
4 g Chuan Xiong
3 g ginseng fibrous roots
6 g Bei Qi
3 g cinnamon
2 dried longans (de-seeded)
6 g Chinese wolfberries (added last)

Seasoning:
1.5 tbsp premium light soy sauce
50 g black glutinous rice wine

Aromatics for dipping sauce:
1 tbsp black sesame oil
1 tbsp diced ginger
1 tbsp diced shallot
1 tbsp diced garlic
1 tbsp premium Shaoxing wine
1 tbsp black glutinous rice wine

Dipping sauce base:
5 tbsp water
2 tbsp light soy sauce
1/2 tbsp dark soy sauce
1/4 tsp salt
1/4 tsp chicken bouillon powder
2 tbsp brown sugar

Method:
1. Remove the innards and the fat near the tail from the duck. Rinse well and chop into pieces.
2. Heat a wok and put in the duck pieces with the skin down. Fry over low heat until golden. Then toss briefly. Set aside.
3. Heat another wok and stir fry the aromatics for ginger soup until fragrant. Put in the duck and sizzle with rice wine. Stir well. Add ginger soup base and bring to the boil. Turn to low heat and cover the lid. Simmer for 1 hour. Add seasoning and stir well.
4. To make the dipping sauce, heat a wok and stir fry the aromatics for dipping sauce until fragrant. Add the dipping sauce base and bring to the boil. Serve on the side with the duck soup.

Tips:
- *Ginger duck soup has a fairly high alcohol content. It shouldn't be seasoned with salt. Otherwise, the soup may taste bitter.*
- *Add Chinese wolfberries into the soup at the last half an hour, otherwise the soup would be sour taste if boiling for a long time.*

Crispy roast goose in Shum Tseng style

Makes 12 servings

Ingredients:
1 black and brown goose (about 3 kg)

Marinade A:
4 tbsp sugar
1 tbsp table salt
1 tsp chicken bouillon powder
1/2 tbsp sesame paste
1/2 tbsp Hoi Sin sauce
1 tbsp oyster sauce
1 tbsp premium Shaoxing wine

Marinade B:
3 whole pods star anise
3 slices ginger
1 sprig coriander
3 shallots

Basting sauce:
120 g white vinegar
48 g maltose
6 g red vinegar

Dipping sauce:
100 g homemade barbecue sauce (see p.10 for method)
3 tbsp plum sauce

Method:
1. Remove the innards, windpipe and fat from the goose. Rinse and drain.
2. Mix marinade A well and brush on the insides of the goose. Put in marinade B ingredients. Close the cuts on the goose with metal skewers.
3. Boil a big pot of water. Dip the duck in the water until the skin tightens and discolours. Remove from water and rinse in cold tap water until the skin feels cool and non-sticky. Hang the goose up to air dry. Brush basting sauce on the skin. Let it air dry again.
4. Preheat an oven to 150°C. Put the goose in a baking tray with the breast side up. Bake for 50 minutes until the skin turns light pink. Cover with aluminium foil and turn the oven down to 100°C. Bake for 30 more minutes until cooked through. Hang the goose up again and remove the metal skewers. Let any liquid drip from insides of the goose.
5. Heat a wok of oil up to 200°C. Deep fry the goose until crispy. Chop into thin slices and save on a serving plate. Serve with homemade barbecue sauce and plum sauce on the side.

Tips:
- *After the goose is grilled, the juices inside are extremely hot. Thus, use care when you remove the metal skewers and let the liquid drip. Don't burn yourself.*
- *After deep-frying the goose, hang it up for 15 minutes before slicing. That would allow time for the juices to return to the flesh, so that the meat would be juicier.*

Marinated chicken in sweet vinegar sauce

Makes 6 servings

Tips:

- *When you make the sweet vinegar sauce, you may put in shelled hard-boiled eggs in the last 5 minutes of cooking time. Bring to the boil and turn off the heat. Just let the eggs soak in the sauce for 3 hours and they are ready to serve.*
- *If you prefer a more robust flavours in your pork trotter and hard-boiled eggs, soak them in hot spiced marinade for 30 minutes before serving.*

Ingredients:
1/2 dressed chicken (about 650 g)
3 slices ginger
15 g spring onion (cut into short lengths)

Sweet vinegar sauce:
1.5 kg Pat Chun brand sweet vinegar
450 g ginger or young ginger
450 g pork trotters (about 6 trotters)

Spiced marinade:
900 g water
900 g light soy sauce
16 g dark soy sauce
400 g brown sugar
40 g Chinese rose wine
2 tbsp five-spice powder
25 g sand ginger roots
4 whole pods star anise

Sweet vinegar sauce:
1. Scrape off the skin of the ginger with a metal spoon. Crush with the flat side of a knife gently.
2. Burn off the hair on the pork trotters. Trim off the nails. Boil water and add ginger and spring onion. Cook the trotters in the water over low heat for 30 minutes. Drain and rinse well.
3. Heat a clay pot. Stir fry ginger with a little of oil until fragrant. Put in pork trotters and sweet vinegar. Bring to the boil again. Cover the lid and cook over low heat for 90 minutes.

Assembly:
1. Dress the chicken and remove innards. Rinse well.
2. Slowly bring the spiced marinade to the boil over low heat. Stir until brown sugar dissolves. Put in the chicken and bring to the boil. Turn to low heat and cook for 5 minutes. Turn off the heat and cover the lid. Leave it in the hot marinade for 30 minutes until cooked through. Drain and set aside.
3. Chop the chicken into pieces. Transfer into a clay pot. Add sweet vinegar sauce and bring to the boil. Serve.

Doggie duck pot

Makes 6 servings

Ingredients:
- 1/2 duck (about 1 kg)
- 300 g lotus root

Aromatics for spiced marinade:
- 1 tbsp oil
- 20 g sliced ginger
- 30 g shallot (sliced)
- 10 g garlic (sliced)
- 3 whole pods star anise
- 1/2 tbsp cloves
- 10 cardamom pods
- 2 tbsp premium Shaoxing wine

Spiced marinade base:
- 900 g water
- 50 g white fermented tofu
- 4 tbsp Unforgettable sauce (see method on p.12)
- 1/2 dried tangerine peel
- 8 sand ginger roots
- 1/2 tsp table salt
- 1/2 tsp chicken bouillon powder
- 1 tbsp light soy sauce
- 1 tbsp dark soy sauce

Thickening glaze:
- 2 tbsp water
- 1 tbsp potato starch

Method:
1. Remove the innards and fat from the duck. Rinse and chop into pieces.
2. Peel and rinse the lotus root. Slice into round discs.
3. Put the duck into a frying pan with the skin side down. Fry over low heat until skin turns golden and the flesh lightly browned. Set aside.
4. Heat oil in a wok and stir fry aromatics for spiced marinade until fragrant. Add spiced marinade base and bring to the boil. Put in the duck and lotus root. Bring to the boil again. Turn to low heat and cover the lid. Simmer for 50 minutes. Stir in the thickening glaze at last. Cook until it thickens. Serve.

Tips:
When you chop up the duck, it's advisable to discard the tail. It's too gamey in taste and is best avoided.

Braised goose with mashed taro in earthenware bowl

Makes 12 servings

Ingredients:
- 1 black and brown goose (about 3.3 kg)
- 24 pieces taro
- 1 sprig spring onion
- 1 sprig coriander

Marinade (to be brushed on the insides of the goose):
- 3 tbsp sugar
- 1 tbsp table salt
- 1/2 tbsp chicken bouillon powder
- 1/2 tbsp five-spice powder
- 2 whole pods star anise
- 3 slices ginger
- 1 sprig spring onion

Spiced marinade base:
- 3.6 kg water
- 5 tbsp table salt
- 1.5 tbsp chicken bouillon powder
- 9 tbsp dark soy sauce
- 6 whole pods star anise
- 12 slices sand ginger
- 9 bay leaves
- 30 g ginger
- 3 tbsp premium Shaoxing wine

Goose bone stock:
- 1 goose head
- 1 goose neck
- 2 goose wings
- 2 goose feet
- 3 slices ginger

Mashed taro:
- 350 g taro (peeled)
- 350 g goose bone stock

Deep-fried aromatics:
- 4 shallots
- 6 slices ginger
- 6 cloves garlic

Mashed taro sauce:
- 4 tbsp cooked goose oil
- 50 g shallot
- 30 g garlic
- 30 g sliced ginger
- 2 tbsp premium Shaoxing wine
- 120 g goose juices (see step 3)
- 200 g goose bone stock
- 700 g mashed taro
- 1 tbsp Chu Hau sauce
- 1 tbsp sesame paste
- 1 tbsp oyster sauce

Tips:
- When you cook the goose at last in the earthenware bowl, do not use high heat. The mashed taro sauce tends to burn and stick to the bottom very easily.
- After making the goose bone stock, transfer all solid ingredients to the spiced marinade base and then boil with the goose meat for enhancing the flavour.
- Close the cut on the goose halfway with a metal skewer, let the heat entering to the cavity and the meat is easily to be cooked. It would also keep the juices inside the goose.

Preparations:

Spiced marinade:
Boil the spiced marinade base. Turn to low heat and cook for 15 minutes. Turn off the heat and let cool.

Mashed taro:
1. To make the goose bone stock, rinse all ingredients. Put them into 1.2 litre of water. Bring to the boil. Turn to low heat and cook for 90 minutes. Strain well. Skim off the oil on the surface. This is the goose bone stock, divide into 350 g and 200 g.
2. To make the mashed taro, slice 350 g of taro into pieces. Steam for 20 minutes. Let cool. Add 350 g of goose bone stock and stir well.

Taro pieces and deep-fried aromatics:
1. To pre-cook the 24 taro pieces, trim them into 1 X 2 inches x 6 mm thick rectangular pieces. Heat oil in a wok. Deep fry taro until golden. Drain and set aside.
2. Heat oil in a wok. Put in the deep-fried aromatics. Fry until golden. Set aside.

Mashed taro sauce:
1. To make cooked goose oil, fry goose fat in a dry wok over low heat until the fat turns into golden cracklings. The oil rendered is cooked goose oil (see photo 1-2 on p.103).
2. Heat up goose oil in a wok. Stir fry shallot, garlic and ginger until golden. Sizzle with premium Shaoxing wine. Add the remaining ingredients. Cook over low heat until it boils. Turn off the heat.

Method:
1. Remove the solid fat from inside the goose (use it to render goose oil). Remove innards. Chop off the feet, wings, neck and head (use it to make stock). Rinse well.
2. Mix the marinade and brush on the insides of the goose evenly. Leave it for 30 minutes. Close the cut on the goose halfway with a metal skewer. Leave the remaining half open.
3. Boil water in a steamer. Put in the goose flat with the breast facing up. Steam for 1 hour till done. Let cool briefly. Pour the goose juices out and save it. Strain the goose juice. Set aside for later use.
4. Chop the goose into quarters. Soak them in the spiced marinade for at least 5 hours. Cut into thin slices.
5. In an earthenware bowl, put spring onion, coriander and the deep-fried aromatics. Then arrange deep-fried taro pieces and goose pieces over them. Pour in the mashed taro sauce. Cook over low heat until it boils. Serve.

Spicy hot pot in Sichuan style

Makes 6 servings

Ingredients of pork bone stock:
600 g pork bones
20 g sliced ginger
1.8 kg water

Aromatics:
1.5 tbsp butter
1 tbsp oil
15 g sliced ginger
15 g spring onion (cut into short lengths)
3 dried chillies
15 g sliced garlic
30 g sliced shallot
2 tbsp spicy bean sauce
3 tbsp Twice-cooked sauce (see p.11 for method)
1 tbsp Sichuan pepper oil
1 tbsp chilli oil

Soup base:
900 g pork bone stock
1 tbsp red chilli powder
1.5 tsp five-spice powder

Seasoning:
1 tsp salt
1/2 tsp chicken bouillon powder
1 tsp oyster sauce

Method:
1. To make pork bone stock, blanch pork bones in boiling water. Transfer into a soup pot. Add sliced ginger and water. Boil over low heat for 3 hours until 900 g of stock remains. Turn off the heat.
2. Stir-fry the aromatics until fragrant. Add soup base ingredients and bring to the boil. Put in the seasoning. Stir well. This is the hot pot base.
3. Put hot pot ingredients on skewers. Dip the skewers into the boiling soup base and cook until done. Serve.

Tips:
- *You may adjust the spiciness of the hot pot base according to your tolerance to spicy food. Adjust the ratio of Twice-cooked sauce, dried chillies and chilli oil freely.*
- *This hot pot works best with ingredients that pick up the spicy soup well, such as fish and meat balls, small pieces of meat, tofu puffs, vegetables and mushrooms.*

Ingredients:
250 g shelled conches
20 g cordyceps flowers
20 g Jinhua ham
20 g sliced ginger
12 Chinese wolfberries
1.2 litre water

Wrappers and ties:
6 cabbage leaves
6 chives

Seasoning:
1/2 tsp salt
1/4 tsp chicken bouillon powder
1/8 tsp ground white pepper

Method:
1. Remove the innards of the conches. Rub 2 tsp of cornstarch and 1 tsp of coarse salt on them to scrub off any dirt. Rinse well with water.
2. Rinse the Jinhua ham. Blanch Jinhua ham and conches in boiling water together.
3. Boil 1.2 litre of water. Put in Jinhua ham, conches, cordyceps flowers and sliced ginger. Cook over low heat for 3 hours. Add Chinese wolfberries and seasoning. Boil for 30 more minutes until 600 g of liquid remains. Turn off the heat. Set aside the solid ingredients with a strainer ladle. Divide the solid ingredients into 6 equal portions.
4. Blanch cabbage leaves and chives in boiling water until soft. Lay flat a cabbage leaf. Put on 1 portion of soup ingredients from step 3. Fold the leaf and tie with a chive into a tight sphere-like package.
5. Put the cabbage rolls back into the soup and cook over low heat for 5 minutes. Serve.

Cabbage rolls filled with conch and cordyceps flowers in soup

Makes 6 servings

Tips:
When you pick the leaves off a cabbage, the stems may snap halfway. You may blanch the whole cabbage until soft before picking the leaves. The leaves can be kept in whole that way.

Chicken and crab hot pot in Sichuan style

Makes 6 servings

Tips:

After stir-frying the Sichuan peppercorns until fragrant, you may transfer them into a tea bag and seal it well. That way, the Sichuan peppercorns won't adhere to the rest of the ingredients and your guests won't have to spit them out.

Ingredients:
- 1 dressed chicken (about 1.35 kg)
- 600 g male mud crab

Marinade:
- 1 tbsp light soy sauce
- 1/2 tbsp dark soy sauce
- 1 tbsp premium Shaoxing wine
- 1 tbsp sugar
- 1/2 tbsp chicken bouillon powder
- 1 tsp five-spice powder

Aromatics for Sichuanese sauce:
- 2 tbsp oil
- 30 g sliced ginger
- 40 g sliced shallot
- 40 g sliced garlic
- 1 sprig spring onion (cut into short lengths)
- 2 tbsp dried chilies
- 2 red chillies
- 2 whole pods star anise
- 2 tbsp Sichuan peppercorns
- 10 sand ginger roots
- 2 tbsp premium Shaoxing wine (added last)

Sichuanese sauce base:
- 300 ml water
- 2 tbsp Lao Gan Ma brand chilli oil
- 1 tbsp oyster sauce
- 1 tbsp Hoi Sin sauce
- 1 tsp spicy bean sauce
- 1 sprig Chinese celery (added last)
- 1 sprig coriander (cut into short lengths, added last)

Method:
1. Remove innards and solid fat from the chicken. Rinse well and chop into pieces. Add marinade and stir well. Leave it for 15 minutes.
2. Drain the chicken. Sear in a frying pan over low heat until golden. Set aside.
3. To dress the crab, put it on a chopping board and chop off its claws first. Flip it over with the belly facing up. Remove the pointy flap. Separate the carapace from the body. Remove the gills. Chop into pieces. Rinse well. Crack the claws slightly with the back of a knife.
4. Heat a clay pot. Stir-fry the aromatics for Sichuanese sauce until fragrant. Put in the chicken pieces and stir well. Add Sichuanese sauce base. Bring to the boil and turn to low heat. Cover the lid and cook for 10 minutes. Put in the crab pieces at last. Toss well. Add Chinese celery and coriander. Cover the lid and cook for 3 more minutes. Serve.

Grilled scallops in duo sauces

Makes 6 servings

Ingredients:
6 live large scallops

Seasoning:
6 tbsp Twice-cooked sauce (see p.11 for method)
6 tbsp Unforgettable sauce (see p.12 for method)
3 tsp grated garlic

Method:
1. Shell the scallops. Keep the half-shells. Use only the scallop meat and the mantle. Rub cornstarch on the scallop meat. Rinse well.
2. Put one scallop and its mantle on each half-shell. Mix seasoning well. Brush it evenly on the scallops.
3. Preheat an oven to 240°C. Bake the scallops on half shells for 4 minutes. Serve hot.

Tips:
- *When you shop for live scallops, pick those with tightly closed shells. If the shells are open or the scallop meat is coming off the shell, those aren't very fresh.*
- *Do not over-bake the scallops. Overcooked scallops tend to be dry and rubbery in texture.*

Pan-fried black cod in black garlic sauce

Makes 4 servings

Tips:

Make sure you de-bone the cod. Otherwise, the bones will stick out in the frying process when the flesh shrinks. That way, the fish cannot be fry properly as the bones stop it from lying flat in the pan.

Ingredients:
   450 g black cod

Marinade:
   240 g water
   2 tsp salt
   1 tsp sugar
   1 tsp chicken bouillon powder
   1 tbsp premium Shaoxing wine
   30 g grated garlic

Aromatics:
   2 tsp oil
   5 g diced ginger
   10 g spring onion (cut into short lengths)
   10 g sliced garlic
   1 tsp rice wine (added last)

Black garlic sauce:
   15 g black garlic (finely diced)
   3 tbsp homemade barbecue sauce (see p.10 for method)
   3 tbsp water

Thickening glaze:
   1/4 tsp cornstarch
   1/2 tsp water

Method:
1. Rinse the black cod. Chop into 6 pieces and de-bone. Add marinade and mix well. Leave it for 25 minutes. Drain well.
2. Heat a wok and stir fry the aromatics in a little oil. Add black garlic sauce ingredients. Bring to the boil over low heat. Stir in thickening glaze. Turn off the heat.
3. Add 2 tbsp of cornstarch to the fish. Mix well. Heat oil in a wok. Fry the fish over low heat until both sides golden. Save on a serving plate. Drizzle with the black garlic sauce from step 2. Serve.

Smoked rose-scented pomfret

Makes 6 servings

Tips:
Do not smoke the fish for too long. Though the wok is heated over low heat only, the fish is constantly being heated up. Over-smoking the fish would dry it out and make it crumbly in texture.

Ingredients:
- 1 pomfret (about 530 g)

Seasoning:
- 300 g water
- 4 tbsp French dried roses
- 1 tbsp red yeast rice
- 50 g maltose
- 30 g sugar
- 30 g light soy sauce
- 10 g dark soy sauce
- 5 g chicken bouillon powder
- 13 g table salt

Vegetable juice:
- 160 g carrot
- 80 g onion
- 60 g celery
- 10 g coriander
- 10 g garlic cloves
- 10 g shallot
- 10 g peeled ginger
- 20 g galangal
- 4 tsp Chinese rose wine

Smoking chips:
- 5 dried French roses
- 1 tbsp flour
- 1 tbsp sugar

Preparation:
1. To make the seasoning, put dried roses and red yeast rice in water. Bring to the boil. Turn to low heat and simmer for 5 minutes. Add the remaining seasoning. Stir until sugar and salt dissolve. Turn off the heat. Let cool.
2. To make the vegetable juice, coarsely chop all ingredients. Put them into a blender. Add seasoning mixture from step 1. Blend until smooth.

Method:
1. Do not cut the fish belly open. Just slice into 6 chunks at an angle. Remove the innards and rinse well. Wipe dry. Soak the fish in seasoned vegetable juice for 10 hours.
2. Preheat an oven to 240°C. Put in the pomfret and grill each side for 6 minutes until cooked through.
3. Mix the smoking chips well. Wrap them in aluminium foil. Put the packet at the centre of a wok. Put a steaming rack on top. Put the fish pieces on the steaming rack. Turn on the heat until smoke comes out of the aluminium foil packet. Cover the lid and turn to low heat. Smoke the pomfret for 3 minutes. Serve.

Fisherman's rice balls in seafood soup

Makes 6 servings

Ingredients:
1 grouper (600 g)
6 marine prawns (about 300 g)
240 g steamed rice
12 spinach leaves
1 tsp grated garlic
10 g coriander (finely chopped)
30 g Chinese celery (finely chopped)
20 g spring onion (finely chopped)

Marinade:
250 g spiced white marinade (see p.13 for method)

Seasoning for spinach:
table salt
chicken bouillon powder

Stock:
6 prawn shells
1 fish bone
20 g sliced ginger
1 tsp ground cumin
1/4 tsp ground white pepper
1.8 kg water

Seasoning for stock:
1/2 tsp table salt
1/4 tsp chicken bouillon powder

Method:
1. Fillet the grouper. Cut into small pieces. Keep the fish bone for making the stock.
2. Shell and devein the prawns. Set aside the shells for making the stock.
3. Heat a wok and add 1 tsp of oil. Stir-fry spinach and grated garlic briefly. Sprinkle with seasoning. Toss well and set aside. Let cool and finely chop them.
4. Pound the steamed rice with pestle and mortar until it resemble sticky dough. Add chopped spinach. Knead to spread evenly. Divide into 12 equal pieces. Roll each into a small ball.
5. Add marinade to the grouper fillet and prawns. Mix and leave them for 20 minutes. Rinse off the marinade with water. Add 2 tbsp of cornstarch. Stir well.
6. Bring the stock to the boil. Turn to low heat and simmer for 1 hour until 900 g of liquid remains. Strian well. Add seasoning for stock. Stir well.
7. Put rice balls into the stock. Cook over low heat for 5 minutes. Add grouper fillet and prawns. Bring to the boil and cook for 2 minutes. Turn off the heat. Sprinkle with coriander, Chinese celery and spring onion. Serve.

Tips:
- *Heat enough the day-old steamed rice if you used before pounding it.*
- *Before rolling the rice balls, you may grease your hands with some oil so that the rice won't stick to them.*

Ingredients:

6 razor clams (about 500 g)
6 tsp Twice-cooked sauce (see p.11 for method)
3 tsp homemade barbecue sauce (see p.10 for method)
6 tsp finely chopped spring onion

Method:

1. Cut each razor clams in half along the length. Remove the innards and dirt. Rinse and drain.
2. Arrange the razor clams on a steaming plate. Spread 1 tsp of Twice-cooked sauce on each razor clam. Leave them for 3 minutes.
3. Boil water in a steamer. Steam the razor clams for 3 minutes. Remove from heat. Sprinkle with spring onion. Drizzle with homemade barbecue sauce. Serve.

Steamed razor clams with Twice-cooked sauce

Makes 6 servings

Tips:

Before dressing the razor clams, soak them in lightly salted water for 2 to 3 hours. This step allows time for razor clams to spit the sand out. No one wants sand in his food I suppose.

Ingredients:

600 g geoducks (about 5 geoducks)
45 g Chinese celery (diced)
15 g coriander (diced)
15 g spring onion (finely chopped)

Carrot puree:

900 g water
200 g carrot
60 g sliced ginger
7 perilla leaves
5 bay leaves

Seasoning:

1.5 tsp table salt
1 tsp light soy sauce
1 tsp chicken bouillon powder

Geoduck in perilla carrot puree

Makes 5 servings

Tips:

Do not overcook the geoducks as they turn rubbery and tough easily.

Method:

1. Cut open the geoduck shells with a paring knife. Remove the brownish black innards. Rinse well.
2. Put 900 g of water and carrot into a blender. Blend into puree. Transfer into a pot. Bring to the boil. Add sliced ginger, perilla leaves and bay leaves. Cover the lid and cook over low heat for 5 minutes. Sprinkle with seasoning. Mix well.
3. Put the geoducks into the puree. Cook over low heat for 2 minutes. Add Chinese celery, coriander and spring onion. Serve.

Roast suckling pig stuffed with bi-colour glutinous rice

Makes 8 servings

Ingredients:
1 dressed suckling pig (about 2.4 kg)

Marinade:
3 tbsp sugar
2 tbsp table salt
1/2 tbsp chicken bouillon powder
1/2 tsp five-spice powder

Basting sauce:
4 tbsp basting sauce A
(see p.13 for method)

Glutinous rice:
900 g glutinous rice
1.2 litre water
300 g purple sweet potatoes
3 tbsp turmeric
150 g preserved pork sausage
150 g preserved pork belly

Seasoning for glutinous rice:
3.5 tsp light soy sauce
3.5 tsp oyster sauce
3.5 tsp Maggi's seasoning
1/2 tsp dark soy sauce
1/2 tsp sesame oil
4.5 tbsp hot drinking water

Tips:
- *When you cut open the spine of the pig or debone it, use care not to puncture through the skin.*
- *After brushing on basting sauce, you don't need to air dry the pig. Just put it straight into an oven.*
- *When you bake the pig, it's just to cook it through. You don't need to brown it at all. If you find any spot starting to brown, you should cover it up with aluminium foil. Otherwise, that spot would burn when deep-fried at last.*
- *When you pour hot oil in the pig, any bubble or trapped air should be pierced through. Otherwise, the air may expand too much and become blisters.*
- *As the pig is deboned before being stuffed, make sure you secure it with fork and metal wires. Otherwise, the stuffed glutinous rice may fall apart.*

Method:

Bi-colour rice:

1. Peel purple sweet potato and dice finely. Put it in a blender and add 600 g of water. Blend until fine. Pour the puree into a tray and add 450 g of glutinous rice. Let the rice soak for 10 hours (or overnight). This is the purple rice. Rinse the starch and drain well.

2. To make the turmeric rice, add 3 tbsp of turmeric to 600 g of water. Mix well and put in a tray. Add 450 g of glutinous rice. Leave it for 10 hours (or overnight). This is the turmeric rice. Drain well.

3. Steam purple rice and turmeric rice separately for 30 minutes. Rinse in warm water to loosen the grains. Drain again.

4. Steam the preserved pork sausage and preserved pork belly for 30 minutes. Finely dice it. Divide into 2 equal portions.

5. Heat a wok over low heat. Put in 1 portion of the preserved sausage and pork belly. Fry briefly to render fat. Put in the purple rice and half of the seasoning. Toss well and set aside. Repeat this step with the turmeric rice.

Stuffing the pig and grilling:

1. Lay the suckling pig flat on a counter with the skin side down. Cut open the spine from the tail towards the head without cutting all the way through. Remove and discard the brain.

2. Lay the pig flat with the skin side down. Slowly remove the first 4 pairs of ribs near the head. Then remove the shoulder bones. Trim off some meat in the thickest part. Make light crisscross incisions on the skin.

3. Boil a pot of water. Put in the pig and cook over low heat for 2 minutes. Drain and soak the pig in water to cool it off. Drain again.

4. Lay the pig flat with the skin side down. Rub marinade on the insides of the pig and leave it for 10 minutes. Skewer the pig on the metal fork and stretch it with aluminum roasted pig planks (see photo 1-5 on p.132).

5. Rinse the pig skin with water. Wipe dry. Spread basting sauce all over.

6. Preheat an oven to 240°C. Put in the pig with the skin side up. Bake with the door ajar for 25 minutes. Turn to 120°C and bake for 45 minutes. Remove from oven and let cool.

7. Debone the pig, except the bones in the head and thighs (see photo 7-8 on p.133).

8. Stuff the pig with the purple and turmeric rice. Wrap the rice in aluminium foil. Closing the both sides of the pig together so that the rice isn't exposed. Stitch the seam on the pig's belly with metal wire (see photo 9-12 on p.134).

9. Put the pig on the metal fork. Secure with metal skewers and metal wire (see photo 13-14 on p.134).

10. Heat oil in a wok until smoking hot. Pour hot oil on the pig's skin until crispy. Then burn the skin with a kitchen torch until bubbly and dry. Remove the aluminium foil and cut into pieces. Serve (see photo 15-19 on p.135).

Deep-fried grouper with eggplant filling

Makes 6 servings

Ingredients:
1 dressed grouper (about 500 g)
1/2 eggplant
1 tbsp premium Shaoxing wine

Marinade:
300 g spiced white marinade (see p.13 for method)

Aromatics:
1 tbsp diced ginger
1 tbsp finely chopped spring onion
1 tbsp grated garlic

Seasoning:
200 g water
1.5 tbsp ground dried plaice
1 tbsp Twice-cooked sauce (see p.11 for method)
1 tsp spicy bean sauce
1 tsp oyster sauce
1 tsp dark soy sauce
1/8 tsp salt

Thickening glaze:
1 tsp cornstarch
2 tsp water

Method:
1. Scale the fish and remove the innards. Remove the backbone and spine bone by cutting between the backbone and the flesh. Rinse well. Add marinade and leave it for 20 minutes. Wipe dry.
2. Heat oil in a wok. Insert the eggplant into the space where the backbone formerly is. Put the fish in the oil and hold it down. Deep-fry over low heat for 5 minutes until the fish resemble a round deep bowl. Drain and put on the serving plate.
3. Take the eggplant out of the fish. Trim off all fried parts. Dice 120 g of eggplant into 1-inch cubes. Deep-fry in oil until golden. Drain.
4. Heat a wok and add 1 tbsp of oil. Stir-fry aromatics until fragrant. Sizzle with 1 tbsp of Shaoxing wine and toss further. Add seasoning and eggplant. Bring to the boil. Turn to low heat and simmer until the eggplant is tender. Stir in thickening glaze. Transfer the eggplant into the fried fish. Serve.

Tips:
When you deep-fry the grouper, make sure you hold it down with a spatula so that the eggplant touches the bottom of the wok. Otherwise, the fish may shrink and separate with the eggplant.

Crispy sesame chicken

Makes 6 servings

Ingredients:
- 1 dressed chicken (about 13.5 kg)
- 2 tbsp premium Shaoxing wine
- 3 tbsp sesames

Marinade:
- 2 tbsp table salt
- 1 tbsp chicken bouillon powder
- 1 tbsp five-spice powder

Basting sauce:
- 1 egg white

Dipping sauce:
- Japanese sesame salad dressing

Method:
1. Remove solid fat and innards from the chicken. Rinse and drain well.
2. Rub premium Shaoxing wine all over the skin of the chicken. Mix the marinade and rub all over. Leave it for 30 minutes.
3. Put the chicken into a steamer with the breast facing up. Steam for 40 minutes. Rinse the skin with cold water. Hang the chicken up to drip any liquid.
4. Brush egg white on the chicken skin. Sprinkle with sesames. Hang to air dry.
5. Heat oil up to 200°C. Deep-fry the chicken until crispy. Chop into pieces. Serve with the dipping sauce on the side.

Tips:
After you brush egg white on the chicken skin, sprinkle with sesames immediately so that they adhere securely. Then wait for 15 minutes for the egg white to bond with sesames before blowing dry with a hair dryer (if you want).

Deep-fried smoky squabs scented with Longjing and chrysanthemums

Makes 6 servings

Tips:
Wait till the spiced marinade is cooled completely before adding premium Shaoxing wine and liquid smoke. Otherwise, the alcohol and wine flavour and smoky smell will evaporate due to the heat.

Ingredients:
2 large squabs

Spiced marinade:
1.2 litre water
1 tbsp five-spice powder
15 g sliced ginger
6 bay leaves
15 g dried chrysanthemums
8 g Longjing tea leaves (added last)

Seasoning:
2 tbsp table salt
1 tbsp brown sugar
1 tbsp chicken bouillon powder
2 tbsp premium Shaoxing wine
125 g liquid smoke

Basting sauce:
4 tbsp basting sauce A
(see p.13 for method)

Method:
1. Remove solid fat and innards from the squabs. Rinse well. Cut off their feet and set aside.
2. To make the marinade, put dried chrysanthemums into a muslin bag and tie well. Boil all spiced marinade ingredients in a pot (except Longjing tea leaves). Turn to low heat and cook for 5 minutes. Turn off the heat. Put in Longjing tea leaves and soak for 2 minutes. Strain the marinade. Add table salt, brown sugar and chicken bouillon powder. Stir until they dissolve. Let cool completely and add premium Shaoxing wine and liquid smoke.
3. Boil a pot of water. Put in the squabs and bring to the boil again. Turn to low heat and cook for 5 minutes. Turn off the heat and cover the lid. Leave them in the water for 30 minutes till cooked through. Transfer into the spiced marinade from step 2. Put them in fridge. Leave them to soak 12 hours for flavours to infuse.
4. Drain the squabs. Brush basting sauce over them. Hang them to air dry. Deep fry in hot oil until crispy. Serve.

Ingredients:

6 prawns (shelled), 1/2 piece pork cheek
1/2 smoked duck breast, 50 g purple sweet potato
6 baby asparaguses, 1/2 onion, 1 sprig coriander
6 white cabbage leaves,
12 baby Napa cabbage leaves, 10 tbsp honey
6 slices pork belly (about 2 cm thick each)

Marinade:

5 tbsp sugar, 1 tbsp table salt
1/2 tbsp chicken bouillon powder
1 tbsp Hoi Sin sauce, 1 tbsp sesame paste
2 tbsp oyster sauce, 1 tbsp premium Shaoxing wine
1 egg, 1 tbsp grated garlic
1 tbsp finely chopped shallot, 2 tbsp grated ginger

Preparations:

1. Rinse the pork cheek. Add marinade and mix well. Leave it for 30 minutes. Bake in an oven at 240°C for 8 minutes on each side. Turn the oven down to 120°C. Bake for 15 minutes.
2. Thaw the smoked duck breast. Bake in an oven at 180°C for 10 minutes.
3. Blanch white cabbage leaves in water over low heat for 5 minutes until soft.

Method:

1. Marinate the prawns with the marinade. Leave them for 10 minutes.
2. Cut grilled pork cheek, smoked duck breast, purple sweet potato, asparaguses, onion and coriander into 3-inch long strips (about the thickness of a chopstick). Divide them evenly into 6 portions.
3. Lay flat a white cabbage leaf. Put a baby Napa cabbage leaf on top. Put one prawn and one portion of the meat filling from step 2 over it. Top with another baby Napa cabbage leaf. Fold the white cabbage leaf upward to wrap all filling securely.
4. Brush marinade evenly on a slice of pork belly. Wrap the cabbage roll with the pork belly tightly. Secure the seam with a toothpick. Transfer into a baking tray. Repeat this step with all remaining cabbage rolls.
5. Preheat an oven to 240°C. Put in the cabbage rolls and grill each side for 10 minutes. Turn oven down to 150°C. Bake for 10 more minutes. Brush honey over them. Bake at 240°C until honey bubbles on both side. Serve.

Grilled cabbage rolls with four-treasure filling

Makes 6 servings

Tips:

- *If you can't get sliced pork belly, you can freeze the whole chunk of pork belly. Then remove the skin and slice it thinly yourself.*
- *When you wrap the cabbage rolls in pork belly, do not leave any space between the two. Otherwise, the pork belly is likely to be dried out in the baking process.*

Beef Bak Kut Teh
(Beef bone herbal soup)

Makes 6 servings

Ingredients:
600 g beef ribs

Spiced marinade:
700 g water
3 g Dang Gui
5 g ginseng roots
3 g Bei Qi
1.5 g cinnamon
2 whole pods star anise
6 g Yu Zhu
5 g Dang Shen
2 dried longans (shelled and deseeded)
3 whole heads garlic
3 tbsp light soy sauce
1.5 tbsp oyster sauce

Dipping sauce:
2 tbsp homemade barbecue sauce (see p.10 for method)

Method:
1. Trim off any solid fat on the beef ribs. Blanch in boiling water for 2 minutes. Rinse in cold water. Set aside.
2. Boil the spiced marinade. Put in the beef ribs and bring to the boil. Turn to low heat and cover the lid. Cook for 90 minutes until the ribs are tender. Serve.

Tips:
- *Do not cook the beef ribs over high heat. Otherwise, the meat will turn crumbly and dry. It has to be cooked over low heat for prolonged period of time to be tender and juicy.*
- *You may also use pork ribs instead of beef for the authentic Bak Kut Teh.*

Ingredients:

- 1.2 kg pork belly
- 6 tbsp potato starch
- 6 tbsp honey
- 4 tbsp homemade barbecue sauce (see p.10 for method)

Marinade:

- 5 tbsp sugar
- 1 tbsp table salt
- 1 tsp chicken bouillon powder
- 2 tbsp Hoi Sin sauce
- 2 tbsp oyster sauce
- 1/2 tbsp sesame paste
- 1/2 tsp light soy sauce
- 1/2 tsp dark soy sauce
- 3 tbsp finely chopped preserved black olives
- 1 tbsp fermented black beans (crushed)
- 1 tsp Chinese rose wine
- 1 tbsp grated garlic
- 1/2 tbsp finely chopped shallot

Method:

1. Skin the pork belly. Trim the meat so that it is a rectangular strip about 1.5 inch thick. Mix 3 tbsp of potato starch with some water. Spread the slurry over the pork belly. Leave it for 4 hours in the fridge.
2. Rinse the pork belly. Brush potato starch slurry over the pork belly again. Put it into the marinade. Leave it for 80 minutes. Set aside.
3. Preheat an oven to 240°C. Grill the pork belly on both sides until lightly caramelized. Turn to 100°C and bake for 30 minutes. Remove from oven.
4. Brush honey on the pork belly. Bake at 240°C until honey bubbles. Turn to 100°C and bake for 20 minutes. Remove from oven and brush with honey again.
5. Before serving, slice the pork belly and drizzle with homemade barbecue sauce. Serve.

Grilled pork belly in preserved black olive sauce

Makes 6 servings

Tips:
Try to ask the butcher to give you a piece of pork belly that doesn't taste gamey. Otherwise, the pork won't taste as good.

Marinated stuffed pork tripe

Makes 6 servings

Ingredients:
1 pork tripe
600 g pork belly
100 g preserved pork belly
50 g peanuts
50 g lotus seeds

Aromatics:
2 tbsp oil
40 g sliced ginger
60 g sliced shallot
60 g sliced garlic
1 tbsp premium Shaoxing wine (added last)

Spiced marinade:
2 kg water
100 g light soy sauce
100 g brown sugar
2 tbsp Lao Gan Ma brand chilli oil
1 tbsp spicy bean sauce
1/4 tsp table salt
4 tbsp dark soy sauce
45 g fermented red tarocurd
3 tbsp Hoi Sin sauce
12 white cardamom pods
3 g cassia bark
2 whole pods star anise
6 sand ginger roots
1 tsp white peppercorns
2 tsp five-spice powder

Thickening glaze A:
1/2 tsp potato starch
1 tsp water

Thickening glaze B:
1 tsp potato starch
2 tsp water

Tips:

Pork tripe must be cleaned thoroughly before cooked. Otherwise, its strong offal taste may interfere with the flavours of the marinade.

Method:

1. Soak peanuts and lotus seeds in water for 2 hours. Drain.
2. Trim off any solid fat on the outside of the pork tripe. Flip it inside out. Divide the coarse salt, potato starch and white vinegar into 3 each portions. Rub the insides of the pork tripe thoroughly with one portion each time. Rinse after scrubbing. Repeat three times to remove the slime and unpleasant smell inside the pork tripe thoroughly. Trim off the yellow membrane. Rinse again and flip the pork inside out again.
3. Cut pork belly into rectangular strips about the thickness of your finger. Cook in boiling water over low heat for 10 minutes. Drain.
4. Stir fry the aromatics in a little oil until fragrant. Add marinade and bring to the boil. Put in pork belly, peanuts and lotus seeds. Bring to the boil and turn to low heat. Cover the lid and cook for 90 minutes.
5. Steam preserved pork belly in a steamer for 30 minutes. Slice thinly.
6. With a hemp string, tie the smaller opening of the pork tripe tightly. Then stuff the pork tripe with pork belly, preserved pork belly, peanuts and lotus seeds via the bigger opening. Heat up 100 g of spiced marinade and stir in thickening glaze A. Bring to the boil. Pour it into the pork tripe. Tie a hemp string around the opening tightly.
7. Put the stuffed pork tripe into the spiced marinade. Bring to the boil. Cover the lid and turn to low heat. Cook for 1 hour and flip the pork tripe upside down every 15 minutes, so that it picks up the marinade flavour evenly.
8. Transfer the stuffed pork tripe onto a serving plate. Cut open with scissors. Meanwhile, heat up 160 g of spiced marinade. Stir in thickening glaze B. Bring to the boil. Drizzle the glaze over the pork tripe and filling. Serve.

燒味 • 傳承滋味　*Authentic Flavours of Barbecue Meat*

作者　Author
陳永瀚　Philip Chen

策劃 / 編輯　Project Editor
Karen Kan

攝影　Photographer
Imagine Union

美術設計　Design
Charlotte Chau

出版者　Publisher
Forms Kitchen

香港鰂魚涌英皇道 1065 號　Room 1305, Eastern Centre, 1065 King's Road,
東達中心 1305 室　Quarry Bay, Hong Kong
電話　Tel　2564 7511
傳真　Fax　2565 5539
電郵　Email　info@wanlibk.com
網址　Web Site　http//www.formspub.com
　　　　http//www.facebook.com/formspub

瀏覽網站

會員申請

發行者　Distributor
香港聯合書刊物流有限公司　SUP Publishing Logistics (HK) Ltd.
香港新界大埔汀麗路 36 號　3/F., C&C Building, 36 Ting Lai Road,
中華商務印刷大廈 3 字樓　Tai Po, N.T., Hong Kong
電話　Tel　2150 2100
傳真　Fax　2407 3062
電郵　Email　info@suplogistics.com.hk

承印者　Printer
百樂門印刷有限公司　Paramount Printing Company Limited

出版日期　Publishing Date
二零一六年七月第一次印刷　First print in July 2016